Contents

INTRODUCTION

People have gazed at the starry skies with wonder and awe ever since the first sparks of human consciousness flickered in the minds of our distant ancestors. Astronomy is the oldest of all the physical sciences; records of celestial events go back many thousands of years to the beginnings of the earliest cultures. Celestial cycles – the movement of the Sun, the phases of the Moon and the annual seasons with their grand parade of stars and constellations – gave humans a practical means of timekeeping; later, the great seafaring civilizations of antiquity learned to navigate by the stars. Things have come full circle; the most accurate clocks we know of are to be found in the regular flashes from exotic stars called 'pulsars', while above us in orbit GPS satellites enable navigators to pinpoint their position on Earth to within metres.

Examples of three celestial objects that can be enjoyed with the unaided eye, binoculars or a telescope: the Orion Nebula (left), a stellar birthplace; Mars (centre); and the Pleiades (right), a cluster of young hot stars.

SEEING THE NIGHT SKIES

There's a great deal to see in the night sky. Whether you use binoculars or a telescope, or even if you have no optical aid at all, there are enough celestial sights to keep anyone enthralled for a lifetime. Stars and constellations are permanent fixtures; others, like the planets, move against the celestial backdrop and appear to change over time. A few phenomena, such as meteors and eclipses, are fleeting but spectacular.

So much can be seen in the night skies without optical aid. It's fascinating to spend time learning the layout of the skies, the positions of the main constellations and the names of the brightest stars – essential, even, if you would like to take your enjoyment of astronomy to a higher level. There are advantages to living in an urban area, where sky-glow caused by city lights drowns out all but the brighter stars. Although just a few hundred stars might be seen with the unaided eye from a dark enclave in a city centre, the night sky appears less crowded and the patterns of the main constellations are easier to trace. Under a dark rural sky where several thousand stars can be seen, the heavens can appear so congested with stars that even experienced astronomers can become a little disorientated.

USING BINOCULARS

Eminently portable, with wide fields of view and low magnification, binoculars give unlimited freedom to roam the night skies – they are the ideal starter instruments for anyone who is new to stargazing. Binoculars gather more light than the eyes alone and reveal many of the night sky's hidden treasures; they also give the startling (although illusory) impression of three dimensions in space.

The colours of stars are especially noticeable through binoculars, and many star clusters, nebulae and galaxies, as well as countless glorious starfields, can be viewed with their aid.

Comparisons of views through different binoculars: the Moon, as seen through 7x30s (left), 12x50s (centre) and 25x100s (right).

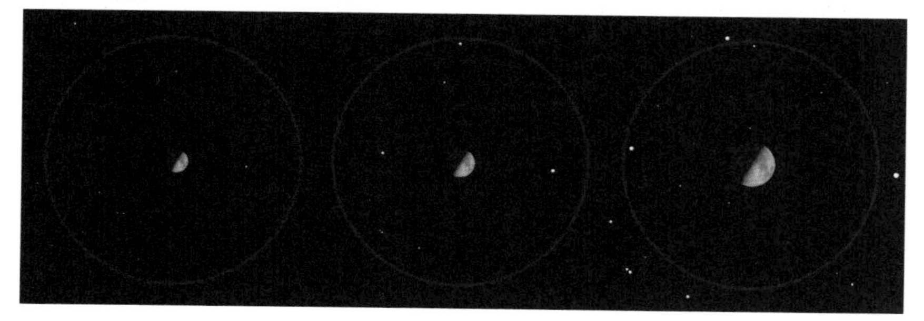

...SEEING THE NIGHT SKIES

The power of binoculars is identified by two figures – one denoting their magnification and the other the size of their objective lenses. For example, a 7x30 binocular – the smallest practically useful for stargazing – magnifies seven times and has 30mm lenses. Even 7x30s can show several hundred thousand stars and numerous deep-sky objects, while more than a million stars and thousands of deep-sky wonders can be seen through a large 20x80 binocular.

USING A TELESCOPE

With their greater light gathering ability, telescopes deliver detailed, magnified views of the night skies. Innumerable deep-sky treasures can be viewed in detail and wonderful structure is visible on the Moon and planets. Viewed through a telescope, the stars themselves appear brighter but they are so far away that they remain pinpoints of light, regardless of the magnification used.

Spectacular images from the Hubble Space Telescope and a bewildering assortment of satellites and space probes show a magnificently pin-sharp, multicoloured Universe.

It's easy to see why some people's expectations are incredibly high when they put their eye to the telescope eyepiece for the first time.

While a great many celestial sights are truly amazing visually, in practical terms much of what the night skies has to offer its Earth-based viewers is relatively faint and requires a degree of informed appreciation. Cosmic colours are generally on the subtle side and the eye has to get used to its own limitations, as well as the limitations of the instrument used and those imposed by the local environment.

The enjoyment, however, is in the self-discovery of the night skies, learning what each celestial object is and where it's located in the Universe.
The realization that your eyes are receiving rays of light from a distant object that may have set out before you were born, before Rome was built or even before humans evolved on Earth, is truly awe-inspiring.

Comparisons of views through different sized telescopes at the same magnification: the Trifid Nebula, as seen through a 60mm refractor (left), a 200mm Schmidt- Cassegrain (centre) and a 500mm reflector (right).

Sunrise over the Parthenon temple on the Acropolis hill in Athens. Built in the 5th century bc, the Parthenon is astronomically aligned with the rising of the Pleiades star cluster in Taurus. *(page 5)*

CELESTIAL GRAPHICS – CHARTING THE STARS

In keeping with the deep human desire to find some sort of order in the cosmos, patterns of stars in the night skies were assembled into constellations – creatures, objects and symbols outlined by prominent stars in a join-the-dots fashion. Constellations reflected the mythology and lifestyle of each culture that imagined them. Such celestial picture books had more than poetic purposes; agricultural communities used their rising, culmination and setting for timekeeping, while navigators and explorers found them to be useful signposts in the sky.

From the Sumerian and Babylonian civilizations of Bronze Age Mesopotamia arose the origins of the patterns of constellations we recognize today; they defined the ecliptic (the yearly path of the Sun), the 12 divine constellations of the Zodiac along the ecliptic (through which the Moon and planets appear to move), and numerous other constellations that referred to animals and agriculture.

These ancient constellations were later incorporated into works by Eudoxus of Cnidus (c.408–347bc), who devised a complete system of the Universe, envisaging Earth at the centre of a series of nested transparent crystal planets. Using naked-eye sighting devices, Hipparchus (c.190–120bc) made detailed observations of star positions, enabling him to create the first known star catalogue. It featured 48 classical constellations and around 850 stars whose position on the celestial sphere was pinpointed according to a system of celestial coordinates.

A magnitude scale devised by Hipparchus denoted each star's apparent brightness; the brightest 20 stars were classed as being of the first magnitude, followed by the next brightest which were second magnitude, and so on, down to the faintest stars, which he classed as sixth magnitude. A similar scale of star brightness is used today, although each division between magnitudes corresponds to a precise jump in brightness by a factor of 2.512 (gauged by photo-electric means).

Some time later, Claudius Ptolemy (c.90–168ad) compiled the **Almagest**, which used and expanded upon Hipparchus's work by producing a definitive atlas of the stars – 1022 of them, contained within the 48 classical constellations, themselves grouped into northern, Zodiacal and southern constellations.

Much of our knowledge of Greek philosophy – including the work of Hipparchus and Ptolemy – comes from ancient texts that were translated, copied and preserved by Arab scholars in Baghdad during the European Dark Ages. Abd al-Rahman al-Sufi (Azophi, 903–86), one of the greatest Arabic astronomers, produced **The Book of the Fixed Stars**, his own version of Ptolemy's star catalogue in which many of the stars were given Arabic names. Many of these names (albeit in modified form) remain in use to this day.

Andromeda (Andromeda, a princess)
Aquarius* (the Water-Carrier)
Aquila (the Eagle)
Ara (the Altar)
Argo Navis** (the Argo, a ship)
Aries* (the Ram)
Auriga (the Charioteer)
Boötes (the Herdsman)
Cancer* (the Crab)
Canis Major (the Great Dog)
Canis Minor (the Lesser Dog)
Capricornus* (the Goat)
Cassiopeia (Cassiopeia, a queen)
Centaurus (the Centaur)
Cepheus (Cepheus, a king)
Cetus (the Whale)
Corona Australis (the Southern Crown)

Corona Borealis (the Northern Crown)
Corvus (the Crow)
Crater (the Cup)
Cygnus (the Swan)
Delphinus (the Dolphin)
Draco (the Dragon)
Equuleus (the Little Horse)
Eridanus (the River Eridanus)
Gemini* (the Twins)
Hercules (Hercules, a hero)
Hydra (the Water Snake)
Leo* (the Lion)
Lepus (the Hare)
Libra* (the Scales)
Lupus (the Wolf)
Lyra (the Lyre)
Ophiuchus (the Serpent Holder)

Orion (Orion)
Pegasus (Pegasus, the winged horse)
Perseus (Perseus, a hero)
Pisces* (the Fishes)
Piscis Austrinus (the Southern Fish)
Sagitta (the Arrow)
Sagittarius* (the Archer, a centaur)
Scorpius* (the Scorpion)
Serpens (the Serpent)
Taurus* (the Bull)
Triangulum (the Triangle)
Ursa Major (the Great Bear)
Ursa Minor (the Little Bear)
Virgo* (the Virgin, a goddess)

*A Zodiacal constellation
**This was later split into three constellations – Carina (the Keel), Puppis (the Poop Deck) and Vela (the Sails).

REDEFINING THE UNIVERSE

Ancient Greek ideas eventually found their way back into the European arena during the High Middle Ages, as the works preserved by Arab scholars were translated into Latin. In the 16th century, Europe saw an explosion of scientific and astronomical enquiry when ancient explanations of the Universe were questioned and found wanting.

In his book De Revolutionibus Orbium Coelestium (**On the Revolutions of the Celestial Spheres**) Nicolaus Koppernik (Copernicus, 1473–1543) promoted the heliocentric theory – a model that places the Sun, not the Earth, at the centre of the Universe. Tycho Brahe (1546–1601), the last and greatest observer of the pre-telescopic era, made precise measurements of the stars and the movements of the planets using naked-eye quadrants and cross-staffs.

Using Tycho's data, his student Johannes Kepler (1571–1630) placed Copernicus's heliocentric theory on a firm scientific footing. Johann Bayer (1572–1625) used Tycho's star positions to produce much of the **Uranometria**, the first star atlas to cover the entire celestial sphere; the far southern stars, uncharted by classical scholars, were mapped according to the catalogue of the navigator Pieter Keyser (c.1540–96).

Uranometria's 51 charts contain more than 2,000 stars, and 12 new constellations were allocated to the deep southern skies.

Importantly, **Uranometria** introduced the system of identifying the brighter stars in each constellation (down to the sixth magnitude) with letters of the Greek alphabet – Alpha being the brightest, Beta the second brightest, and so on.

In some large constellations the Greek letters ran out, so Bayer used Roman letters, starting with a uppercase A followed by lowercase b, c, d, and so on. The system, devised before the telescope was invented, was neither precise nor perfect, but for all its idiosyncrasies it is retained to this day in much the same form as it originated.

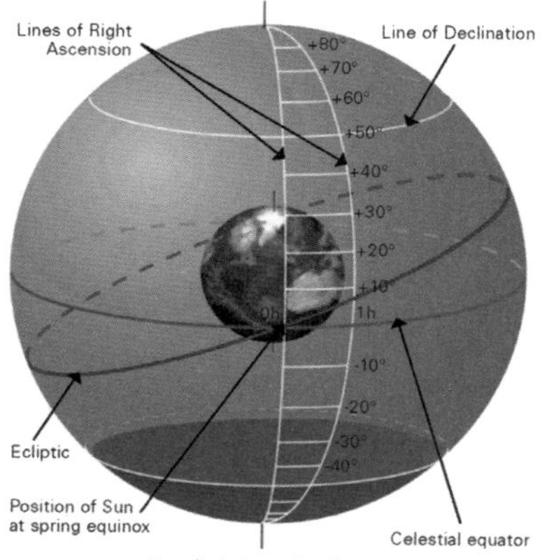

THE GREEK ALPHABET (WITH SYMBOLS)

Alpha α / Beta β / Gamma γ / Delta δ / Epsilon ε
Zeta ζ / Eta η / Theta θ / Iota ι / Kappa κ
Lambda λ / Mu μ / Nu ν / Xi ξ / Omicron ο
Pi π / Rho ρ /Sigma σ / Tau τ / Upsilon υ
Phi φ /Chi χ/ Psi ψ/ Omega ω

The celestial sphere, showing the north and south celestial poles, the celestial equator and the ecliptic. The shaded area around the south celestial pole is the region uncharted by the star-mappers of the ancient world.

TELESCOPIC REVELATIONS

Undeniable proof of a radically different layout of the Universe came with the invention of the telescope in the early 17th century. Galileo Galilei (1564–1642) discovered four satellites orbiting Jupiter, and Venus's phases showed that it was a globe in orbit around the Sun. It became increasingly obvious that Earth was just another planet with a satellite, orbiting the Sun between Venus and Mars.

Galileo discovered that the faint band of the Milky Way was made up of countless stars that were only visible through the telescope. It was reasoned that if the stars themselves were like the Sun – but so distant that they appeared as points of light – then perhaps the Sun wasn't so special. Instead of lying at the hub of the Universe, the Sun was found to be just one of a broader mass of stars making up the Milky Way.

Great advances were made in our understanding of the Universe through telescopic observation during the 17th, 18th and 19th centuries. As telescopes grew larger, familiar objects became better known, and new objects loomed into view from deeper, darker depths of the cosmos. Telescopic surveys of the skies charted the stars and catalogued deep-sky objects – star clusters and faint misty patches known as nebulae. From his private observatory, Johannes Hewelke (Hevelius, 1611–87) made accurate measurements of star positions and produced the most advanced star atlas of its time, the *Uranographia*. He devised a number of new constellations to lie among the traditional ancient Greek constellations. Then, using observations made from his own private observatory in England, John Bevis (1693–1771) compiled *Atlas Celeste*, an 18th century star atlas containing elaborately engraved star charts.

The 18th century finally saw the beginnings of the proper charting of the deep southern skies – a portion of the celestial sphere that had never been surveyed accurately before. Nicolas Lacaille (1713–62) set up an observatory at the Cape of Good Hope, from where he catalogued nearly 10,000 southern stars and 42 deepsky objects. Lacaille's *Coelum Australe Stelliferum* introduced 14 new constel lations in and around Bayer's deep southern constella-tions, all of which became accepted by the astronomical community.

Charles Messier (1730–1817) compiled a list of 110 deep-sky objects. Incorporating all sorts of objects that appear 'fuzzy' through a small telescope, Messier's list is an eclectic mix of star clusters, nebulae and galaxies. It proved so useful that it is still referred to today by stargazers.

Using telescopes that he had made himself, William Herschel (1738–1822) surveyed the skies and recorded hundreds of double stars and nebulae. After discovering Uranus in one of his routine star-sweeps, Herschel established himself as the world's most prolific astronomer and went on to make many discoveries about the Solar System. William's sister, Caroline Herschel (1750–1848) was also a prolific observer of the skies, while his son, John Herschel (1792–1871) surveyed the skies of the southern hemisphere, recording hundreds of previously unseen double stars and nebulae.

As burgeoning European empires extended their reach around the globe, the increasing importance of accurate navigation by the stars led to the founding of national observatories, such as the Paris Observatory, France (1671), the Royal Greenwich Observatory in England (1675) and Germany's Berlin Observatory (1700). Telescopes were used to note the exact time that stars transited the meridian (appeared due south), enabling coordinates to be measured precisely.

From Greenwich the first Astronomer Royal, John Flamsteed (1646–1719) catalogued 2,935 stars and produced the most accurate celestial atlas of its day, the *Atlas Coelestis*; this was updated by Nicolas Fortin (1750–1841) in Paris, producing the smaller, more popular *Atlas Fortin-Flamsteed*, which had an artistic makeover and a number of newly discovered nebulae. In Berlin, Johann Bode (1747–1826) produced the *Vorstellung der Gestirne* (intended for amateur astronomers) and the Uranographia, which contains more than 17,000 stars and deep-sky objects and is widely regarded as the greatest pictorial star atlas of all time. By the late 19th century, photography had advanced to such a state that it was possible to chart the skies on photographic plates.

An international consortium established the *Carte du Ciel*, a project intended to chart the entire sky photographically using a number of identical astrographic telescopes set up around the world. It was an enormous project – more than 22,000 photographic plates were taken between 1881 and 1950 – and the task of physically measuring star positions down to the eleventh magnitude and with an accuracy of 0.5 arcseconds on each plate was incredibly laborious. The project was superseded in the 20th century using the very large wide-angle 48-inch Schmidt telescope at Mount Palomar in California; completed in 1958, the Palomar Sky Survey covers the sky from the north celestial pole to a declination of -30 degrees and shows stars down to the twenty-second magnitude on average (a million times fainter than those visible with the naked eye).

With improvements in optics and photography came wave after wave of deep space discovery. Catalogues of newly discovered deep-sky objects were created to keep a tally, notable among which were John Dreyer's *New General Catalogue* (NGC, 1888) containing nearly 8,000 objects, and its appendices, the *Index Catalogue* (IC, 1895) with more than 5,000 objects. Many hundreds of NGC and IC objects are visible through an average-sized amateur tele-scope and they remain essential lists for today's astronomer.

The late 20th century saw the introduction of orbiting observatories high above Earth's turbulent atmosphere. Foremost among these, in terms of mapping the skies, was the European Space Agency's Hipparcos (High Precision Parallax Collecting Satellite). Between 1989 and 1993 Hipparcos precisely measured the positions of celestial objects, leading to the production of the *Tycho-2 Catalogue*, which contains 2.5 million stars. Data for stars nearby in our Galaxy has allowed astronomers to determine accurately the actual motion of stars through space and to gauge their distance from us using the parallax effect resulting from Earth's orbit around the Sun.

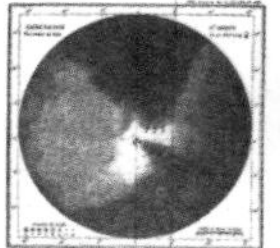
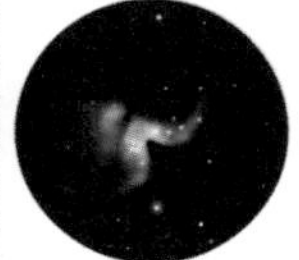

Charles Messier's drawing of the Orion Nebula (M42) compared with the author's observation of the same object.

The constellation of Leo, showing all of its NGC objects – most of them are distant galaxies.

LIVES OF THE STARS

Even though the stars are at unimaginable distances from us, astronomers know a great deal about them and their life cycles. A star's mass is the overriding factor in determining how big it is and how brightly it shines throughout its life, how it develops during its lifetime and how long it lives.

All stars begin their lives in a gravitationally contracting cloud of interstellar dust and gas – mainly hydrogen and helium – within a galaxy. These clouds are so dense and deep as to be virtually opaque to visible light, and some of them can be seen silhouetted against parts of the Milky Way.

Numerous stellar birthplaces may appear within each contracting zone of interstellar dust and gas. It may take around ten million years from the first stages of collapse to the appearance of an embryonic star, a region of unstoppable gravitational collapse called a protostar.

STELLAR IGNITION

Dust and gas attracted by the protostar's gravity produces ever- rising heat and pressure at its core.
Temperatures eventually become high enough to trigger thermonuclear fusion, where two hydrogen atoms combine at high speed to produce one helium atom, giving off a blast of energy, which ignites the star.

Much of the dust and gas surrounding the young star is blown away by the strong stellar wind, but any remaining material in orbit bathes in its new sun's light and energy. Astronomers have been able to detect and image such disks of gas and dust around newborn stars: known as proplyds (protoplanetary disks), they are solar systems in the making. Many beautiful proplyds have been imaged in the Orion Nebula, and among notable naked-eye stars with proplyds are Beta Pictoris, 51 Ophiuchi, Fomalhaut, Vega and Zeta Leporis.

All stars are not born equal. The amount of time a star spends fusing hydrogen into helium – living as a Main Sequence star – depends on its mass. The larger its mass, the hotter the star; its energy production is faster as its consumption of hydrogen fuel takes place at a faster rate, so it enjoys less time on the Main Sequence.

Perhaps the most well-known dark nebula is the Horsehead Nebula in Orion – a remarkably shaped projection of the edge of a larger molecular cloud, which is silhouetted against a glowing red gas cloud in the vicinity of the star Sigma Orionis. 1,500 light years away, the Horsehead Nebula is around ten light years in length. Inset, a closer Hubble Space Telescope view, where a young star, still swaddled in its nursery, can be seen shining at top left.

An 'average' star like the Sun, now around 4.7 billion years old, will spend another five billion years on the Main Sequence before entering its old age. Huge stars with 50 times the Sun's mass will spend just a million years on the Main Sequence before using up all their hydrogen fuel. The smallest stars on the Main Sequence – brown and red dwarfs with a mass less than one-third of the Sun – have convective interiors with hydrogen and helium mixed throughout. This allows them to burn up a far greater proportion of their hydrogen fuel than more massive stars, giving them exceedingly long lives of many tens of billions of years – longer than the current age of the Universe. As a consequence, no red dwarf in the late stages of its life has ever been observed.

A fantastic assortment of proplyds, imaged in the Orion Nebula by the Hubble Space Telescope.

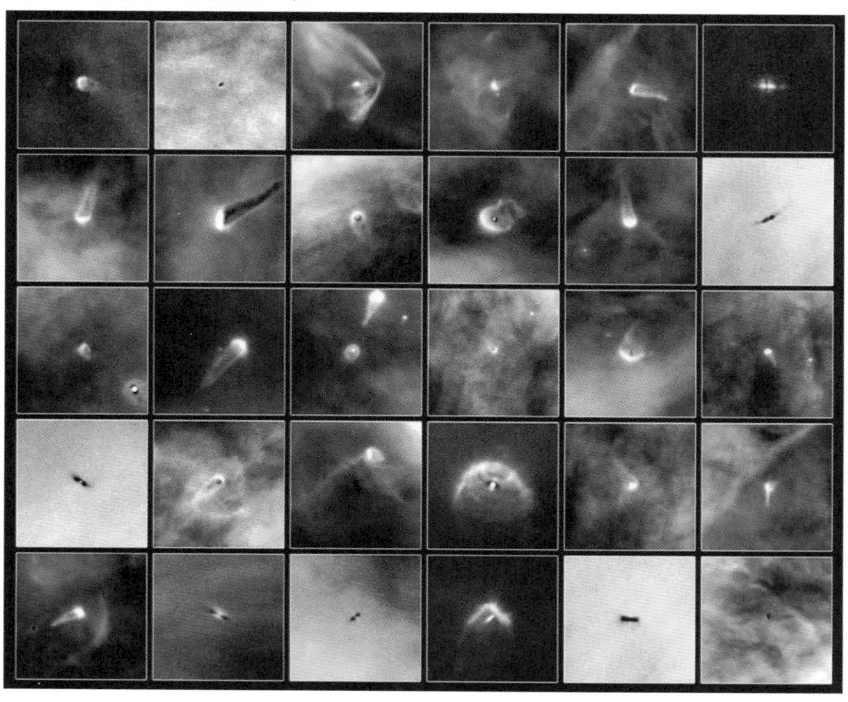

STAR COLOUR AND TEMPERATURE

A star's colour gives us an idea of its surface temperature. Redder stars are cooler, while bluer stars are hotter. Our own yellow Sun has a surface temperature of around 6,000 °C (10,800 °F). Betelgeuse (Alpha Orionis) – an old red supergiant in the final stages of its life, 1,000 times the size of the Sun and 600 light years away – has a surface temperature of around 2,000 °C (3,600 °F), which is about one-third of the Sun's temperature.

It was once maintained that the composition of stars would forever remain a mystery. This was proved wrong when the spectroscope – a special piece of equipment used to split light into a spectrum for analysis – clearly revealed chemical elements in the stars.

A multitude of thin dark lines in stellar spectra, known as absorption lines, are like the fingerprints of the elements. Spectroscopy allows the directional speed of stars to be determined by their 'Doppler shift', a line-of-sight effect that displaces absorption lines towards the red or blue end of the spectrum; the degree to which light is shifted allows its relative speed to be gauged. Redshifted light is produced by a star moving away from the observer, while an approaching star will show a blueshift.
The spectral types are identified by seven main groups – O, B, A F, G, K and M – from hot O-type blue stars, through G-type stars like the Sun, to cool M-type red stars.

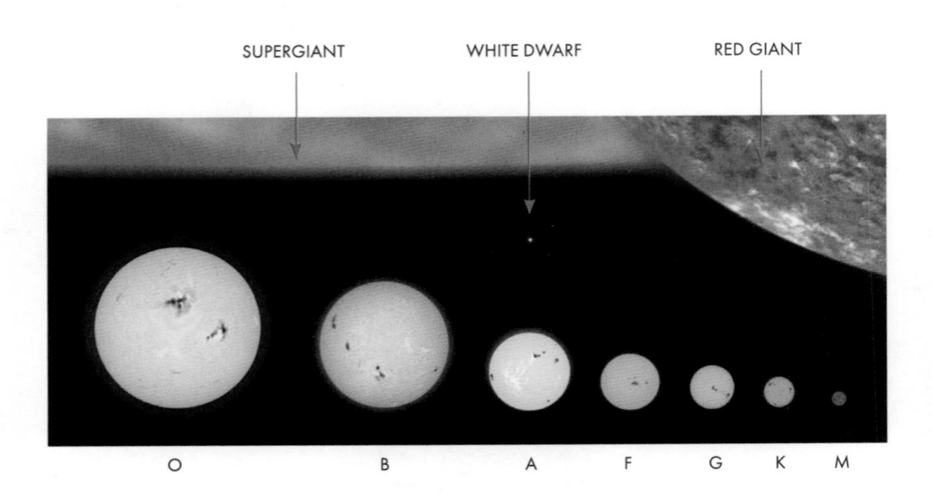

Stars come in a vast range of sizes. Main Sequence stars, from hot O-types, through G-types like the Sun, to M-type red dwarves, compared to off-sequence stars like tiny white dwarf remnants and huge old red giants and supergiants.

EXOPLANETS

It was once thought impossible that planets around other stars (exoplanets) could be detected as stars are so incredibly distant that any planets would appear so close to them as to be unresolvable as individual objects. Moreover, as planets shine only by reflected light, they were considered far too faint to detect, their light being drowned out by starlight. Thanks to ultra-sensitive measurements using the latest technology, the age-old dream of finding exoplanets has now finally been realized.

One technique for detecting exoplanets measures minute variations in a star's velocity (its Doppler shift) caused by the gravitational tug of orbiting planets; the closer and more massive the planet to its parent star, the larger the observed effect. This technique can only tell us the minimum size of the exoplanet, its orbital period and distance from the star. Another detection technique measures variations in a star's brightness caused when, from our viewpoint, a planet passes in front of it; silhouetted against the star, the transiting planet causes a tiny but detectable drop in brightness. This technique reveals a great deal more, including the mass and radius of both the star and planet. Although only a small proportion of stars and their planetary systems are aligned in our line of sight, the transit technique allows astronomers to gauge an exoplanet's gravity and density, and to speculate reasonably about its composition and surface conditions.

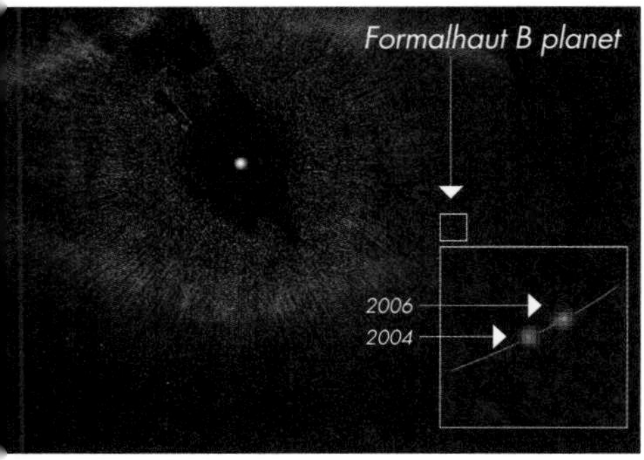

It is now thought that a large proportion of stars have planetary systems, including around half of all Sun-like stars. In our Galaxy alone there are likely to be many tens of billions of planets. In 1995 the star 51 Pegasi became the first Sun-like star known to have an exoplanet – one with half the mass of Jupiter and orbiting at a blistering pace of once every 4.2 days. Since then, hundreds of exoplanets have been discovered in a wide variety of exoplanetary systems.

Exoplanets have been discovered using both Earth-based telescopes and orbiting observatories, notably NASA's Kepler Mission, which was launched in 2009 specifically to discover Earth-like exoplanets in our galactic neighbourhood. Once in orbit, NASA's James Webb Space Telescope will be able to follow up Kepler's amazing discoveries and measure the chemical composition of transiting exoplanets' atmospheres with great accuracy. This is an important step in finding out whether our own planet – orbiting in a 'warm' zone allowing liquid water to exist on its surface and just the right size to hold on to a life-friendly atmosphere – is rare among the stars, or whether conditions amenable to the development of life exist in abundance in the Universe.

Hubble Space Telescope image of the vast dust ring around the bright young star Fomalhaut in Piscis Austrinus, showing the embedded exoplanet Fomalhaut B, about three times the mass of Jupiter.

WHEN THE FUEL RUNS OUT

A star eventually uses up the hydrogen fuel at its core; energy production falls and the core's temperature and pressure decreases. In response, the core contracts slightly, producing a sharp rise in temperature which in turn ignites the hydrogen shell surrounding the core – a zone that was previously too cool to undergo fusion reactions. At this point the star leaves the Main Sequence; as it expands, the star's increased surface area makes it appear brighter, while its surface becomes cooler and redder. The star has become a red giant.

As the star's core contracts yet further and pressure increases, temperatures rise high enough to begin burning the remaining helium within the core, fusing it into carbon. As the core periodically contracts, the star's outer atmosphere is puffed into space as rings or shells of material; they are known as planetary nebulae because some resemble ghostly planets when viewed through a telescope. At their centre lies the highly compressed million-degree stellar remnant – an Earth-sized object called a white dwarf, so dense that a thimble full of its stuff would weigh a metric ton. Hydrogen gas within planetary nebulae glows as it is ionized by ultraviolet radiation emitted by the white dwarf.

But planetary nebulae are short-lived – they are only visible for less than 100,000 years, as they expand and fade.

Around 1,000 planetary nebulae populate our part of the Galaxy. Although they shine for only a brief moment on the cosmic stage, planetary nebulae are among the most beautiful objects in the Universe. Although they have all been formed in the same way, these puffs of gas from dying stars vary enormously in appearance. Easily the largest and brightest of these, the Dumbbell Nebula in Vulpecula, is visible through binoculars; it has two brightly glowing lobes, giving it the appearance of a luminous apple core. Another lovely bright planetary nebula, the Ring Nebula (M57) in Lyra, is a beautiful telescopic sight – a luminous doughnut whose central white dwarf is just visible through larger instruments.

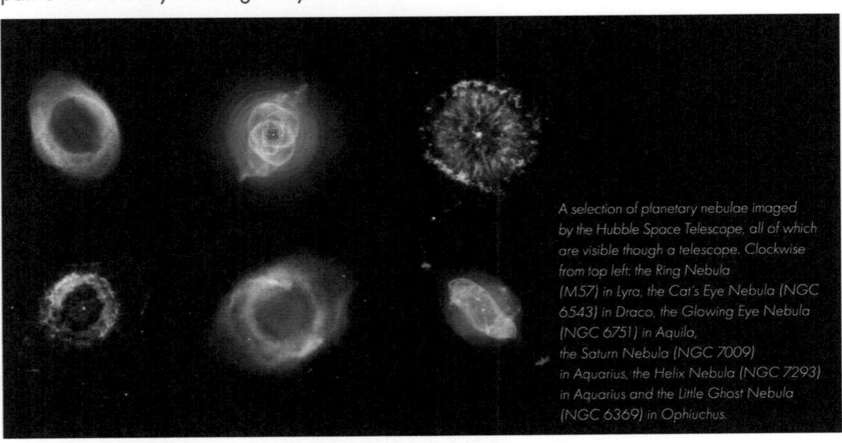

A selection of planetary nebulae imaged by the Hubble Space Telescope, all of which are visible though a telescope. Clockwise from top left: the Ring Nebula (M57) in Lyra, the Cat's Eye Nebula (NGC 6543) in Draco, the Glowing Eye Nebula (NGC 6751) in Aquila, the Saturn Nebula (NGC 7009) in Aquarius, the Helix Nebula (NGC 7293) in Aquarius and the Little Ghost Nebula (NGC 6369) in Ophiuchus.

THE CELESTIAL SPHERE

Since ancient times, the stars were imagined to be arrayed upon the inside of a vast celestial sphere at whose centre lay the Earth; moving between us and the sphere, the Sun, Moon and planets appeared projected against the starry background. Even though we know that the Universe is far different, we continue to use this analogy for positional astronomy because it is easy to work with.

Extensions of Earth's polar axis and equator define the position of the celestial poles and equator, and like a terrestrial globe the celestial sphere has a simple system of coordinates.

Parallel lines of declination measured up to +90 degrees (north) and -90 degrees (south) from the celestial equator correspond to terrestrial latitude, while the celestial equivalent to longitude is represented by great circles of right ascension (RA), measured from 0 to 24 hours around the celestial equator. Each degree of declination and hour of RA is split into 60 arcminutes and each arcminute is further divided into 60 arcseconds.

Our annual orbit around the Sun produces another important line on the celestial sphere – the ecliptic – a line traced out by the Sun's apparent path among the stars and the 12 traditional Zodiacal constellations all straddle the ecliptic. Because Earth's axis of rotation is tilted by 23.5 degrees – the ecliptic makes an angle of 23.5 degrees with the celestial equator.

Earth and all the major planets orbit the Sun in almost the same plane so they follow paths close to the ecliptic.

The phenomena of eclipses, produced when the Earth, Moon and Sun are in alignment, gives us the word ecliptic.

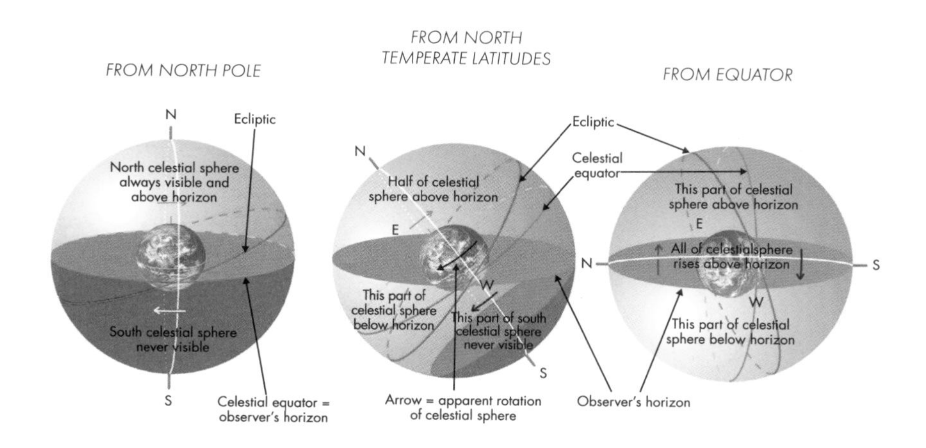

Our location determines how much of the celestial sphere is seen annually. From the poles only half of the celestial equator is ever visible, but all of it can be seen from the equator over time. From temperate locations most of the celestial sphere comes into view except for a region around the opposite celestial pole.

TIME AND SEASON

Rotating on its axis every 24 hours, the Earth orbits the Sun once every 365.25 days – the extra quarter day is added every four years to give us an extra day in February, in the form of a leap year.

While Earth's axis is tilted by 23.5 degrees to the plane of the ecliptic, the tilt is constant with respect to the stars; this gives us the seasons, whose effects are more noticeable further away from the equator.

In December the Earth's north pole is angled away from the Sun, and during the northern winter the Arctic regions are in permanent darkness; meanwhile the southern hemisphere enjoys the height of summer, when the Antarctic is bathed in 24-hour sunlight. Precisely the opposite happens six months later when, during June, the Earth has moved to the other side of its orbit around the Sun. Now the north pole is angled towards the Sun at the height of northern summer, while it is midwinter in the southern hemisphere. These two extremes are called the winter and summer solstices.
Between the solstices are the spring and autumn equinoxes, when neither pole is angled towards the Sun, and all parts of the Earth experience 12 hours of sunlight and 12 hours of darkness. two extremes are called the winter and summer solstices.
Between the solstices are the spring and autumn equinoxes, when neither pole is angled towards the Sun, and all parts of the Earth experience 12 hours of sunlight and 12 hours of darkness.

THE STARS' ANNUAL PARADE

From any particular part of the Earth there is a region surrounding the celestial pole that is always visible, and the extent of this circumpolar region varies with latitude. From either of the Earth's poles, exactly half the sky – everything above the horizon – appears circumpolar. Polaris (Alpha Ursae Minoris, known as the North Star because it is located close to the north celestial pole) is directly overhead at the north pole; as the Earth revolves the stars track parallel to the horizon, neither rising nor setting.
From the latitude of London at 52°N, the north celestial pole is 52 degrees high; everything north of +38 degrees on the celestial sphere is circumpolar, while stars south of this declination appear to rise and set. Nothing south of -38 degrees on the celestial sphere ever rises from London. The same holds true for locations in the southern hemisphere.

From the latitude of Canberra (35°S), all stars south of -55 degrees are circumpolar, while everything north of +35 degrees lies permanently below the horizon.

Viewing non-circumpolar constellations depends largely on the season. The Sun's glare renders the constellations immediately surrounding it practically impossible to view.
As the Sun proceeds along the ecliptic, formerly invisible constellations begin to emerge into the morning skies, reaching their highest above the horizon at midnight around six months later. Constellations then sink gradually into the evening skies, moving ever westward until they are once again lost in the evening afterglow and the glare of the Sun.

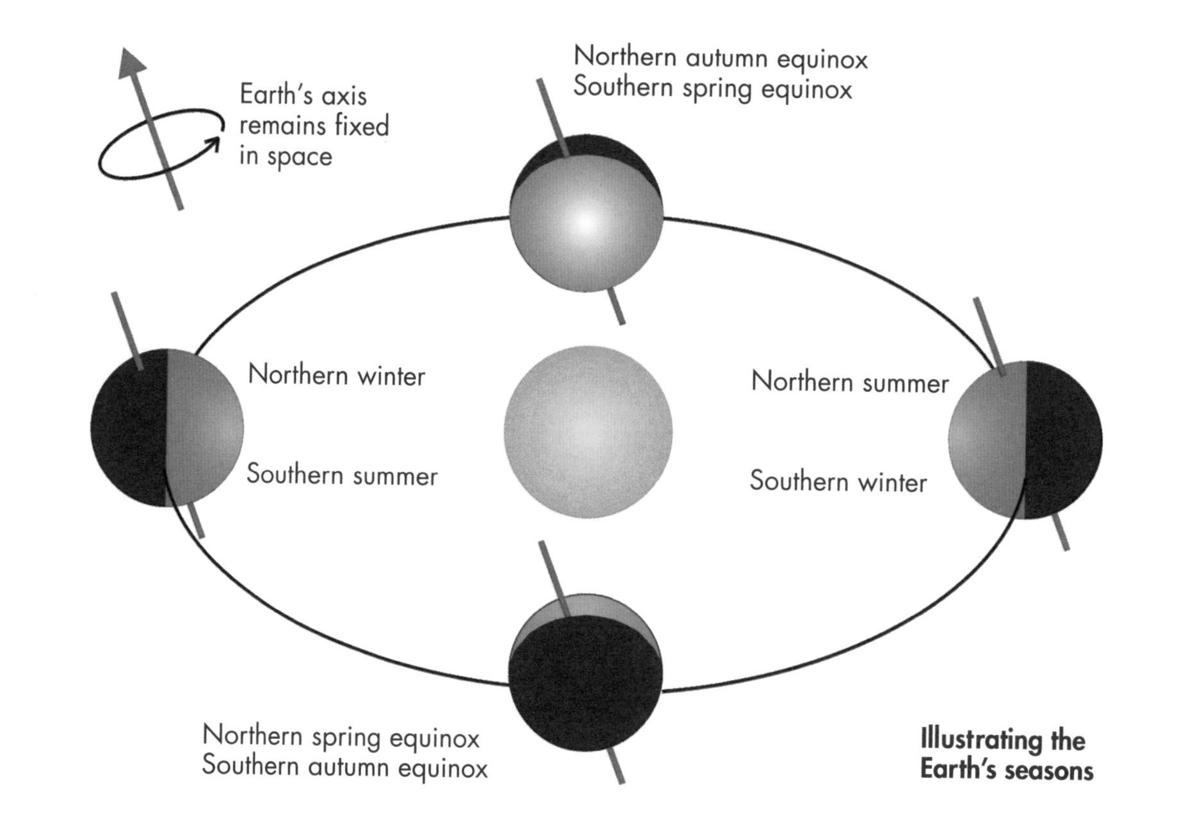

Illustrating the Earth's seasons

THE NORTHERN STARS

In this section we take a look at the main constellations, stars and celestial showpieces of the northern celestial sphere, beginning with constellations around the north celestial pole and then taking a season-by-season view. Most northern constellations are as familiar to today's stargazers as they were to the ancient Greeks.

NORTHERN CIRCUMPOLAR CONSTELLATIONS

From northern temperate climes a number of large constellations – most of them easily identifiable – remain constantly above the horizon throughout the year as they wheel anticlockwise (counter-clockwise) around the north celestial pole.

Polaris, the tip of Ursa Minor's tail, is conveniently located just one degree from the north celestial pole. The star is easily found by tracing a line from Merak and Dubhe in Ursa Major, two stars known as the Pointer.

These two stars are part of the famous Plough, alternatively known as the Big Dipper or sometimes the Saucepan. One of the sky's most easily identifiable asterisms, the Plough forms the tail and hindquarters of Ursa Major. At midnight during the autumn (fall) the Plough is at its lowest, scraping along the northern horizon, while it soars high overhead at springtime. On the other side of the north celestial pole can be found another prominent asterism, the W of Cassiopeia.

The Plough and the W take turns throughout the year in gaining the high celestial ground. Much of the relatively star-sparse region between the Pointers and the W is filled by the large but ill-defined constellation of Camelopardalis, while another large, faint constellation, Lynx, borders Camelopardalis and Ursa Major.

Somewhat easier to trace is the House asterism of Cepheus, which lies between Cassiopeia and the pole.

Between Cepheus and the tail of Ursa Major, skirting around the edge of Ursa Minor, is the sprawling constellation of Draco, covering an area of more than a thousand square degrees. Despite being so large, Draco contains only a few reasonably bright stars ranging between the second and third magnitude. By way of contrast, on the other side of the celestial pole and nestled snugly between Cassiopeia and Auriga, lies beautiful Perseus. For the most part a circumpolar constellation from northern temperate regions, Perseus has a number of bright stars sprinkled along a section of the Milky Way and is one of the loveliest areas to scan with binoculars.

A broad view of the northern circumpolar sky, looking due north (east at right, west at left). The outer circle represents extent of circumpolarity from London (52°N) and the inner circle for stars that are circumpolar from New York (41°N). The ecliptic is also shown at either side (none of it is circumpolar). The chart is relevant for 1 November (4am), 1 December (2am), 1 January (midnight), 1 February (10pm) and 1 March (8pm).

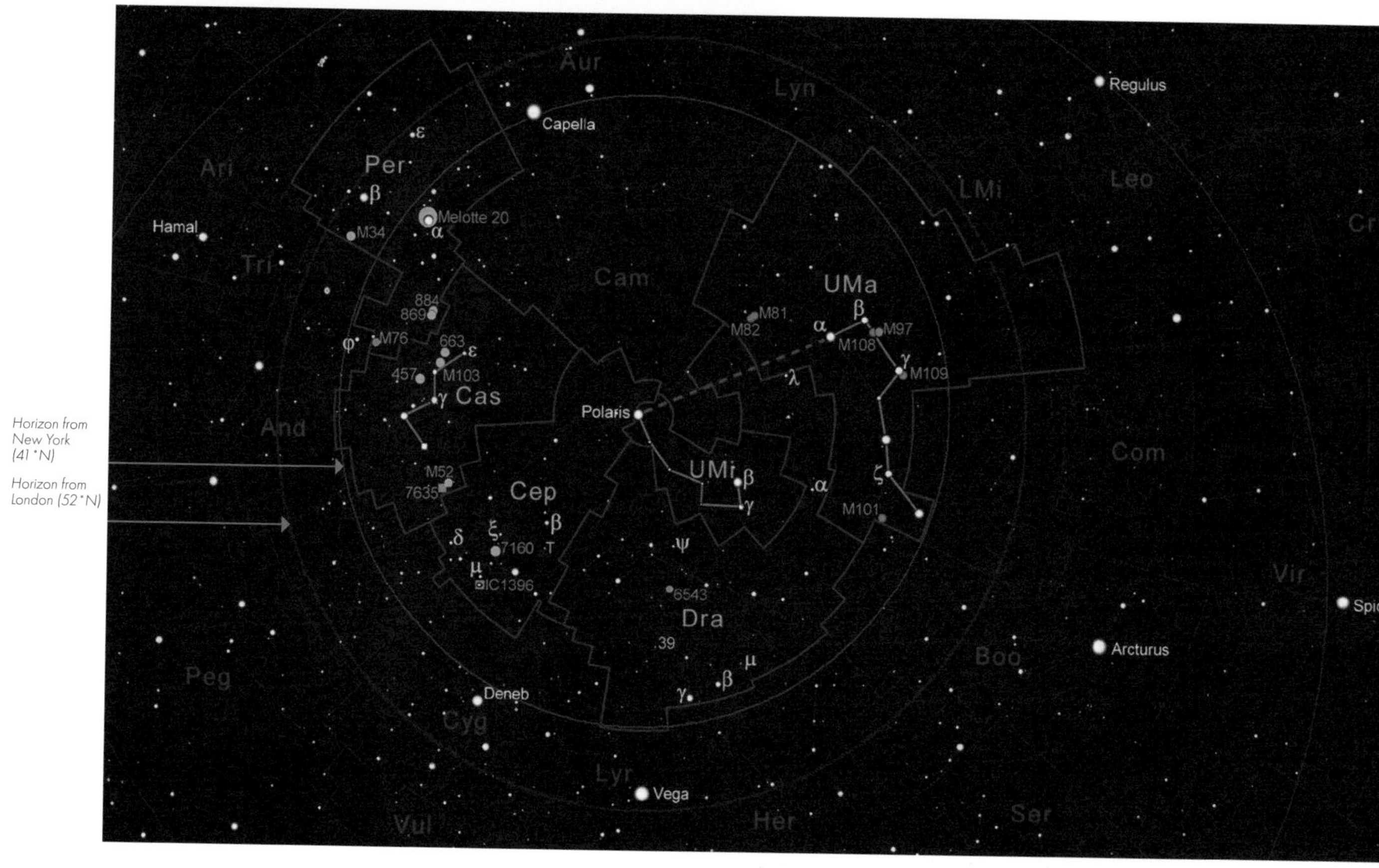
Horizon from
New York
(41°N)
Horizon from
London (52°N)
Polaris
Capella
Hamal
Regulus
Arcturus
Vega
Deneb
Spica
Per
Cas
Cep
Cam
UMa
UMi
Dra
Ari
Tri
And
Peg
Aur
Lyn
LMi
Leo
Crt
Com
Vir
Boo
Ser
Her
Lyr
Cyg
Vul
Melotte 20
M34
884
869
M76
663
457
M103
M52
7635
7160
IC1396
6543
39
M81
M82
M108
M97
M109
M101

URSA MINOR

UMI / URSAE MINORIS

Highest at midnight: early January

Sometimes called the Little Dipper, Ursa Minor is a small but significant constellation incorporating the north celestial pole. Its brightest star, Polaris, is located within just one degree of the pole.

A phenomenon called precession, caused by the slow movement of the Earth's axis, will bring Polaris closest to the pole (within half a degree) at the end of the 21 st century.

Polaris is useful to aim at when aligning an equatorial telescope for a quick observing session.

A telescope roughly aligned with Polaris will keep an object within the field of view of a medium-power eyepiece for a long time before requiring adjustment. A telescope trained on Polaris itself will reveal it to be a double star, with an eighth-magnitude companion. Beta UMi (Kochab) and Gamma UMi (Pherkad), the end stars of the Little Dipper, are sometimes called the Guardians of the Pole.

A keen naked eye will discern a faint star close to Pherkad (about half the Moon's diameter away) – this is an unrelated foreground star.

A little circlet of seventh- and eighth-magnitude stars near Polaris can be seen through binoculars – this is nicknamed the 'engagement ring', Polaris being the bright celestial solitaire.

CEPHEUS

CEP / CEPHEI

Highest at midnight: late February

Cepheus contains several reasonably bright stars. Among the stellar delights of Cepheus is **Beta Cep**, a double star of magnitudes 3.2 and 7.9 that can be resolved through a small telescope.
Beta Cep is a variable star, a blue giant whose brightness fluctuates by around one tenth of a magnitude over a period
of just a few hours. Another variable, which is also a nice double, is **Delta Cep**. The prototype of the Cepheid variables (see Variable stars in Introduction). This is a pulsating yellow supergiant that varies between magnitudes 3.5 and 4.4 in a period of five days nine hours. Delta's companion is a magnitude 6.3 blue star, and the pair is easy to separate through
a small telescope.

Mu Cep is a red supergiant whose striking colour has earned it the nickname of the Garnet Star. Binoculars will show its ruddy hue to good effect. The star is also variable, fluctuating between magnitudes 3.4 and 5.1
over a period of two to two and a half years. Mu Cep is one of the biggest stars visible with the unaided eye – if placed in the position of the Sun, the surface of its enormous bloated sphere would extend almost out to the orbit of Saturn. Xi Cep is a double star comprising a magnitude 4.4 blue star and a magnitude 6.5
orange giant, easily resolvable through a small telescope. T Cep, a red Mira-type variable star, has a period of more than a year. At its maximum brightness the star just pops into the range of the unaided eye,
shining at magnitude 5.2.

The Garnet Star in Cepheus, observed through a 100mm refractor by the author.

Double star Delta Cephei.

An extension of the Milky Way nudges into the southern part of Cepheus, and a couple of lovely open star clusters can be found in this vicinity – IC 1396 and NGC 7160 – both delightful to view through a 150mm telescope. IC 1396 is embedded within a sizeable patch of nebulosity; it is also known as the Elephant's Trunk Nebula because of a prominent dark sinuous dust lane, which is visible on photographs. Large binoculars will reveal IC 1396 as a misty patch. NGC 7160 is a small, compact star cluster; around 30 of its stars are visible through a 200mm telescope, half a dozen of the brighter ones standing out from the rest.

The Elephant's Trunk Nebula in Cepheus, imaged through an 80mm refractor with an astronomical CCD camera (filters used).

URSA MAJOR

UMA / URSAE MAJORIS

Highest at midnight: early March

Seven of the brightest stars within Ursa Major make up an asterism variously called the Plough or the Big Dipper.
While this asterism itself doesn't much look like a bear, a little time spent in tracing the traditional outline comprising the remainder of the constellation's bright stars will convince any stargazer that the ancients who named it had an extremely good eye for form.

The two front stars of the Plough, Alpha UMa (Dubhe) and Beta UMa (Merak) are known as the Pointers, since an imaginary line extending from them leads to Polaris and the north celestial pole. Zeta UMa (Mizar), the second star of the Plough's handle, has a fainter magnitude 4 partner, 80 UMa (Alcor), which is visible with the unaided eye.
Mizar itself is a close double star, with components of magnitudes 2.2 and 3.8, separable with a small telescope.

A pair of galaxies bright enough to be seen through binoculars, Bode's Galaxy (M81) and the Cigar Galaxy (M82) lie in the far north. Just half a degree apart, the pair is visible in the same low-power telescopic field. While M82 is almost edge-on to us, M81 is tilted at less of an angle. Some ten million light years distant, these galaxies are interacting with each other. On the other side of the constellation, the face-on spiral galaxy M101 is visible through binoculars as a circular smudge, and appears mottled through a 200mm telescope.

The Owl Nebula (M97), a faint planetary nebula, appears as a pale disk about twice the diameter of Jupiter through a 150mm telescope. The dark eyes of the owl, so obvious in many images, are rather elusive and require at least a 250mm telescope to discern.

Multiple star Mizar in Ursa Major.

The familiar stars of the Plough in Ursa Major.(page 22)

M108, a bright, sizeable and nearly edge-on galaxy, can fit into the same low-power telescopic field as M97. Bright condensations within M108 can be discerned through a 150mm telescope.

Although the location of the galaxy M109 can be identified fairly easily, being a little more than half a degree east of Gamma UMa, its low surface brightness makes it more of a challenge to observe. A 200mm telescope will reveal its bright elliptical centre along with a superimposed foreground star just to the north of the core, but detail within the spiral arms requires a larger instrument to resolve.

Galaxy M101 in Ursa Major, imaged using a 127mm refractor and astronomical CCD camera.

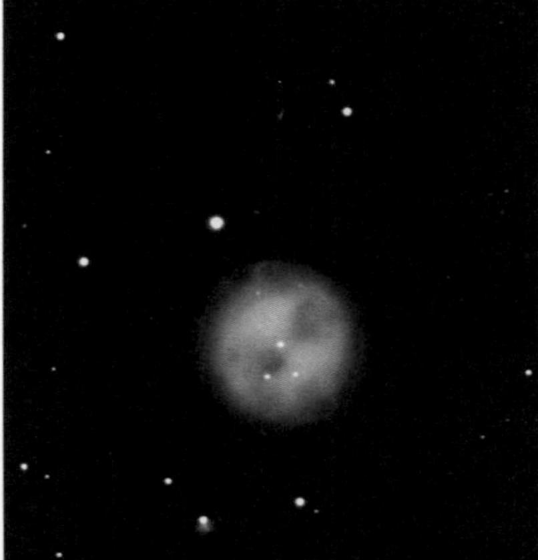

Planetary Nebula M97 in Ursa Major, imaged using a 127mm refractor and astronomical CCD camera.

DRACO

DRA / DRACONIS

Highest at midnight: early July

Despite sprawling across a huge portion of the northern circumpolar region, Draco is not the most prominent of constellations. Its traditional outline can be traced from its head (marked by Beta Dra and Gamma Dra) just north of Hercules, along a winding path to Alpha Dra (Thuban) at the constellation's narrowest part, around to Lambda Dra near Draco's western border. Around the time that the Pyramids were constructed, Thuban (magnitude 3.7) was the brightest star near the north celestial pole; precession will grant it this distinction again in more than 21,000 years' time.

Mu Dra is a close telescopic double star with white components of magnitudes 4.9 and 5.6; the pair is slowly moving apart, and they can now be comfortably resolved with a 100mm telescope, and are a good test for a 60mm telescope. Psi Dra is much easier to resolve; binoculars will reveal the yellow stellar duo of magnitudes 4.6 and 5.8. Binoculars can split the two wide components of 39 Dra (magnitudes 5 and 7.4); a telescope will show the magnitude 8 companion of the brighter star.

The Cat's Eye Nebula (NGC 6543) is a small but bright planetary nebula with a distinct bluish hue. A 150mm telescope will show it as a small ring surrounding an eleventh-magnitude central star.

The Cat's Eye Nebula in Draco, imaged using a 105mm refractor and astronomical CCD camera.

CASSIOPEIA

CAS / CASSIOPEIAE
Highest at midnight: early October

With its prominent five-star W asterism, Cassiopeia is one of the easiest constellations to recognize. Gamma Cas, the central star of the W asterism, is an irregular variable star that fluctuates, at unpredictable intervals, between magnitudes 3 and 1.6. Eta Cas is a nice double star with a magnitude 3.5 yellow primary and a red magnitude 7.5 companion, easily visible through a small telescope.

Cassiopeia is a joy to scan with binoculars, as a bright section of the Milky Way flows across the constellation, engulfing the W. A treasure trove of bright open clusters lies within its boundaries, most of which lie east of the W. Containing around 30 stars, the bright compact cluster of M103 is best seen at higher magnifications.

The Owl Cluster (NGC 457) is a loose assembly of around 100 fairly bright stars arranged in distinct lines; its two brightest stars shine like an owl's eyes. NGC 663 is a beautiful binocular cluster containing around 80 stars. On the far western side of Cassiopeia, the compact Scorpion Cluster (M52) contains around 100 stars, the brightest of which form a splendid S shape. The Bubble Nebula (NGC 7635), a faint diffuse nebula visible through a 200mm telescope, lies just half a degree southwest of M52, so that the two objects can be viewed in the same low-power field of view.

A delightful alignment of stars known as Kemble's Cascade can be found in the circumpolar constellation of Camelopardalis, east of Cassiopeia.

The Bubble Nebula in Cassiopeia, imaged using an 80mm refractor and astronomical CCD camera (filters used).

PERSEUS

PER / PERSEI

Highest at midnight: mid-November

Crossed in the north by the Milky Way, Perseus is a magnificent constellation containing a number of bright stars and open clusters. Near Alpha Per (Mirfak) lies Melotte 20, a large loose star cluster made up of a snaking chain of bright stars;
it is a stunning sight through binoculars and at low magnifications.

Beta Per (Algol) is a famous eclipsing binary. Every 2.87 days it drops from magnitude 2.1 to 3.4, changes easily monitored with the unaided eye. The Spiral Cluster (M34) can just be discerned with the unaided eye some five degrees west of Algol. It contains a number of star chains, with some of its brighter stars paired up.

Eta Per is a nicely coloured double star, easily resolvable through a small telescope, with an orange magnitude 3.8 primary and a blue magnitude 8.5 companion.

Located in the far northwestern corner of Perseus, the Double Cluster (NGC 869 and NGC 884) is one of the most breath-taking sights in the heavens.
These two bright open clusters – each the diameter of the full Moon – lie side by side, and can be glimpsed as a hazy patch with the unaided eye.
A low-power view will accommodate both clusters, revealing hundreds of stars. Several red stars can be discerned near NGC 884, contrasting nicely with the cluster's profusion of blue stars.

Glowing at magnitude 10, the Little Dumb-bell Nebula (M76) is the faintest Messier object (see Telescopic revelations in Introduction). Resembling a tiny apple core, this faint planetary nebula can be seen through a 150mm telescope.
It lies less than one degree north of Phi Per.

Planetary nebula M76 in Perseus, imaged using a 160mm refractor and astronomical CCD camera (filters used).

NORTHERN WINTER STARS
(midnight, 1 January)

Plenty of dazzling stars and bright deep-sky objects are sprinkled about the winter night skies. The Milky Way, dotted with fabulous star clusters, runs from near the zenith to the southern horizon, while the ecliptic slices across the skies, from Virgo in the east, through Leo, Cancer and Gemini to Taurus, Aries and Pisces in the west.

Orion, climbing high above the southern horizon is immediately recognizable, with its bright orange star Betelgeuse, dazzling Rigel and the three bright stars
in a line making up Orion's Belt.

Below, in Orion's Sword Handle, can be seen a fuzzy patch – the glow of the Orion Nebula, one of the most beautiful of all deep-sky objects.

Orion's Belt stars can be used as a handy signpost for finding other stars and constellations. Follow the belt downwards to the left and you'll come
to Canis Major and Sirius, the brightest star in the night skies. Because of its relatively low altitude from mid-northern
latitudes, Sirius often appears to twinkle because of the effects of Earth's atmosphere, producing a multicoloured display of stellar scintillation.

Extend the line of Orion's Belt to the right and you'll find Aldebaran, a bright orange star in Taurus. The V-shaped asterism to which Aldebaran belongs, represented as the horns of a bull on old star charts, is called the Hyades; further to
the west lies the Pleiades, a cluster of bright stars.

Above Orion lies Auriga, its brightest star Capella soaring almost directly overhead. Straddling the Milky Way, Auriga contains a number of beautiful bright open star clusters. Neighbouring Auriga, further along the band of the Milky Way, lies the constellation of Perseus.

To its west lie two of the largest galaxies in our Local Group – M33, the Pinwheel Galaxy and M31, the Great Andromeda Spiral. Both of these nearby galaxies are comparable in size to our own and can be glimpsed with the unaided eye under very dark skies.

High in the south climb the heavenly twins Castor and Pollux in Gemini whose toes dip into the Milky Way. At its heels follows Canis Minor and the bright star Procyon. Meanwhile, as Virgo rises in the east it is preceded by Leo, with its familiar Sickle asterism and bright star Regulus, and the fainter constellation of Cancer. This contains
a large open cluster, the Beehive, easily discerned with the unaided eye between Regulus and Pollux.

Looking towards the northern horizon, the familiar circumpolar stars have completed another quarter-turn about the celestial pole. Ursa Major is in the ascendancy standing on its tail, while Cassiopeia is beginning to sink westwards.

Northern winter sky, looking due south (east at left, west at right) from the horizon to the zenith. The horizon lines for London (52°N) and New York (41°N) are marked, as well as the ecliptic. The chart is relevant for 1 November (4am), 1 December (2am), 1 January (midnight), 1 February (10pm) and 1 March (8pm).

Horizon from New York (41°N)
Horizon from London (52°N)
Arcturus
Spica
Regulus
Alphard
Castor
Pollux
Procyon
Capella
Betelgeuse
Aldebaran
Rigel
Sirius
Canopus
Hamal
CVn
UMa
Lyn
LMi
Com
Leo
Vir
Sex
Crt
Crv
Hya
Cnc
Gem
Aur
Per
Tau
Ori
Mon
CMa
Lep
Col
Pyx
Pup
Ant
Vel
Cen
Pic
Hor
Eri
For
Ari
Tri
M44
M67
M35
M37
M36
M38
M1
M42
M45
2281
2392
1907

TAURUS

TAU / TAURI

Highest at midnight: early December

Brooded over by the red bulls' eye of **Alpha Tau (Aldebaran)**, this large and easily identifiable Zodiacal constellation dominates the skies northwest of Orion. Aldebaran is set against the background of the **Hyades**, a large V-shaped open cluster containing a dozen or more naked-eye stars in an area that can be covered by a clenched fist. Among them, the bright doubles **Theta Tau** and **Kappa Tau** are both separable with a keen unaided eye, but **Sigma Tau** requires some optical aid to split.

In the northwest of Taurus, a handful of the brightest stars within the Pleiades (M45) star cluster is easy to see with the unaided eye.
Telescopes will reveal many dozens of young blue stars, and near the Pleiad 23 Tau (Merope) may be seen a hint of reflecting nebulosity.

Just over one degree north of Zeta Tau, the faintly glowing supernova remnant of the Crab Nebula (M1) requires an 80mm telescope to be seen at all well. Even through large instruments it appears as a grey, rather featureless elliptical patch.

Wispy nebulosity can be seen surrounding the Pleiades star cluster, imaged with a 105mm refractor and astronomical CCD camera.

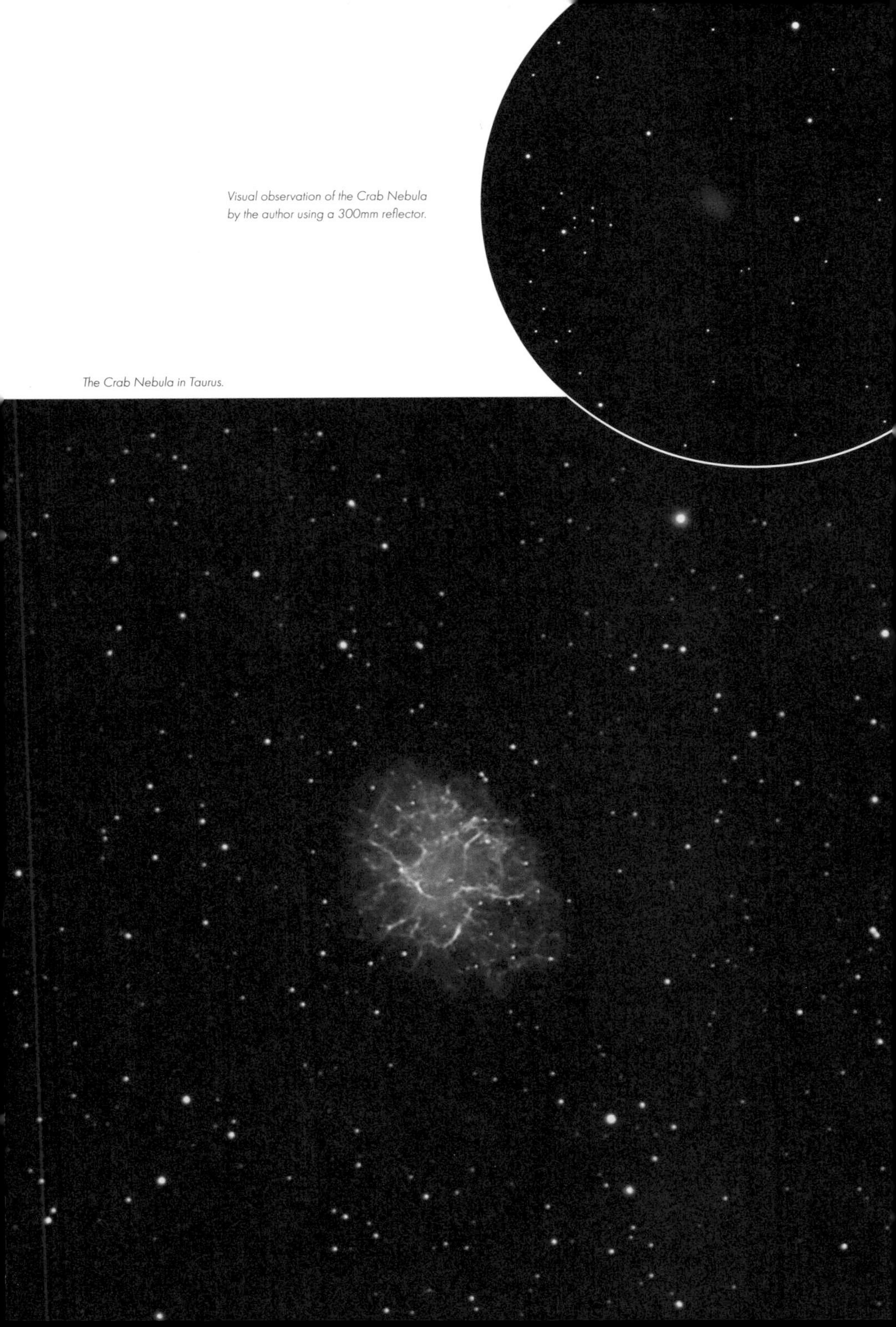

Visual observation of the Crab Nebula by the author using a 300mm reflector.

The Crab Nebula in Taurus.

ORION

ORI / ORIONIS

Highest at midnight: mid-December

Orion is the most magnificent of all the constellations. Straddling the celestial equator, its main shape – made up of the red **Alpha Ori (Betelgeuse)** in the north, the brilliant blue **Beta Ori (Rigel)** in the south and the three dazzling stars of **Orion's Belt** in between – is recognizable to stargazers in both northern and southern hemispheres.

Sigma Ori is a lovely multiple; its main star of magnitude 3.8 has two nearby sixth-magnitude stars, and a more distant triple of eighth-magnitude stars. All six stars can be marvelled at in a high-power field of view.

A short distance south of Orion's Belt, a misty patch can be discerned with the keen unaided eye.

This is the Orion Nebula (M42), one of the biggest and brightest nebulae in the heavens. Considerable structure within the nebula can be seen with binoculars alone, and a small telescope will reveal a glowing greenish mass with delicate wisps, intruded on by a prominent dark lane. Several stars can be seen in and around the nebula, notably the Trapezium (Theta Ori), a bright quadruple star. Larger telescopes will show breathtaking detail within the nebula. Just to its north lies the fainter De Mairan's Nebula (M43), with a single bright star nestling at its centre.

The beautiful Horsehead Nebula in Orion is very challenging to view through even large amateur telescopes, but it is captured here on a CCD image taken through a 127mm refractor (filters used).

AURIGA

AUR / AURIGAE

Highest at midnight: mid-December

Broad, bright and easily recognizable, Auriga rides high overhead on winter nights. Southwest of brilliant **Alpha Aur (Capella)** lies the small triangular asterism of the **Kids.** Its apex star, **Epsilon Aur**, is a white supergiant of magnitude 3 that is eclipsed by an unseen companion every 27 years, when it drops to magnitude 3.8 for a year. Its next eclipse is due to take place around 2037.

Three of the brightest clusters in Auriga – M36, M37 and M38 – are easily visible through binoculars as misty patches in the southern part of the constellation.

Of these, M37 is the biggest and brightest; an orange ninth-magnitude star lies at the centre of around 150 fainter stars, resolvable with a 150mm telescope, with a hint of nebulosity at its centre. M38 is made up of around 100 stars below tenth magnitude, the brightest of which form a startling cross shape. Telescopes will reveal the much smaller and fainter cluster NGC 1907 half a degree south of M38. In eastern Auriga, the cluster NGC 2281 contains a wide scattering of half a dozen fairly bright stars, with more than 20 fainter ones in the background.

Star cluster M36 in Auriga, imaged with a 160mm refractor and astronomical CCD camera (filters used).

GEMINI

♊

GEM / GEMINORUM

Highest at midnight: early January

A sizeable, easily recognized constellation, Gemini is a bright and well-defined constellation lying between Auriga and Canis Minor. Its two main stars, **Alpha Gem (Castor)** and **Beta Gem (Pollux)** are instantly identifiable.
Pollux, slightly the brighter of the pair, has a decidedly orange hue. Castor is a famous multiple star, with its two brightest components (magnitudes 1.9 and 3) resolvable through a good 60mm telescope. Interestingly, both **Eta Gem** and **Zeta Gem** are double stars whose primaries are also variables.

The far western part of Gemini is immersed in the Milky Way. Here, the beautiful open cluster M35 can be glimpsed with the unaided eye, just a couple of degrees northwest of Eta Gem – binoculars show it well.

Spread across an area the size of the full Moon, M35 contains about 80 stars, some of which are arranged into a prominent curved chain.

A couple of degrees southeast of Delta Gem can be found the Eskimo Nebula (NGC 2392), a bright eighth-magnitude planetary nebula that shows up as a greenish blob at low magnifications.
It is surprisingly large as planetary nebulae go, with a diameter equivalent to the apparent size of Jupiter. It has a bright, almost stellar centre.

Castor, the brightest star in Gemini.

Star cluster M35 in Gemini, imaged with a 160mm refractor and astronomical CCD camera (filters used).

CANCER

CNC / CANCRI

Highest at midnight: early February

Cancer is the least prominent of all the 12 Zodiacal constellations. Although it is so faint, the constellation can easily be located to the southeast of the bright stellar twins Castor and Pollux.
Cancer's third- and fourth- magnitude brightest stars form the shape of an inverted Y, which is easy to trace from dark-sky sites.

Small telescopes will split the two main components of Zeta Cnc, widely separated stars of magnitudes 5.2 and 5.8 and a 200mm telescope will reveal that the brighter star has a closer magnitude 6.2 companion.

Iota Cnc is a lovely coloured double, with a yellow magnitude 4 primary and a blue magnitude 6.6 companion, easy to resolve with a small telescope.

Keen eyes will discern a misty patch just north of Delta Cnc and directly in the middle of the rectangular area bounding the constellation. Binoculars will reveal this to be the Beehive Cluster (M44),
a sizeable star swarm covering an area of around ten times that of the full Moon. A very low telescopic magnification will take in the entire cluster. Around 80 stars can be seen, of which the sixth-magnitude
Epsilon Cnc is the brightest.

Less than two degrees west of Alpha CnC lies a much smaller open cluster than the Beehive, the King Cobra (M67).
Despite its large near neighbour, M67 is a major open cluster in its own right, containing 300 stars and appearing as a full Moon-sized oval smudge through binoculars.

A 150mm telescope will resolve the cluster's fainter stars, most of which are below eleventh magnitude.

Star cluster M67 (the King Cobra) in Cancer, imaged with a 105mm refractor and astronomical CCD camera.

NORTHERN SPRING STARS
(midnight, 1 April)

With its milder nights, springtime is eagerly looked forward to by those who prefer milder observing conditions. Having revolved to its highest point in the sky, Ursa Major and the Plough soar high overhead.

Cassiopeia is low above the northern horizon, its W asterism perhaps a little difficult to discern from light-polluted urban areas. More of a challenge to view with the unaided eye, the band of the Milky Way runs at its lowest across the sky, almost parallel to the northern horizon, from Canis Minor in the west through Perseus and Cassiopeia in the north to Aquila in the east. Procyon, Canis Minor's brightest star, is becoming increasingly low in the west and often appears to twinkle because of its low altitude.

Having ascended to near-zenithal heights during the winter months, Capella, the brightest star in Auriga, now slowly sinks towards the northwestern horizon, followed by Casto and Pollux in Gemini. Hard on their heels Leo, with its bright star Regulus, prowls above the southwestern horizon, tentatively eyeing the faint constellation of Cancer.
High in the south, trailing Leo, is the faint constellation of Coma Berenices, notable for its broad cluster of dim stars – the cluster appears as a faint patch of light just beyond most people's naked-eye resolution and is sometimes mistaken for a cloud.

Both Coma Berenices and Virgo to its south contain the richest collection of bright galaxies in the entire sky. Deep-sky observers eagerly explore the Virgo-Coma Cluster – a grouping of more than 1,300 galaxies of varying sizes and shapes lying around 50 million light years away – some of which are bright enough to be seen through binoculars and small telescopes.

Fainter constellations nearby include Libra, Corvus, Crater and Sextans, all underlain by lengthy Hydra whose brightest star, Alphard, lies midway between Regulus and the southwestern horizon. Stretching across the sky from Libra in the southeast, passing Spica and Regulus, the ecliptic reaches to Taurus, low in the northwest.

On the ascendancy, brilliant Lyra in Vega is climbing above the northeastern horizon, followed by Deneb in Cygnus.
Orange Arcturus in Boötes shines brightly in the southeast, while Hercules and the pretty constellation of Corona Borealis are climbing to an increasingly favourable altitude.

Northern spring sky, looking due south (east at left, west at right) from the horizon to the zenith. The horizon lines for London (52°N) and New York (41°N) are marked, as well as the ecliptic. The chart is relevant for 1 February (4am), 1 March (2am), 1 April (midnight), 1 May (10pm) and 1 June (8pm).

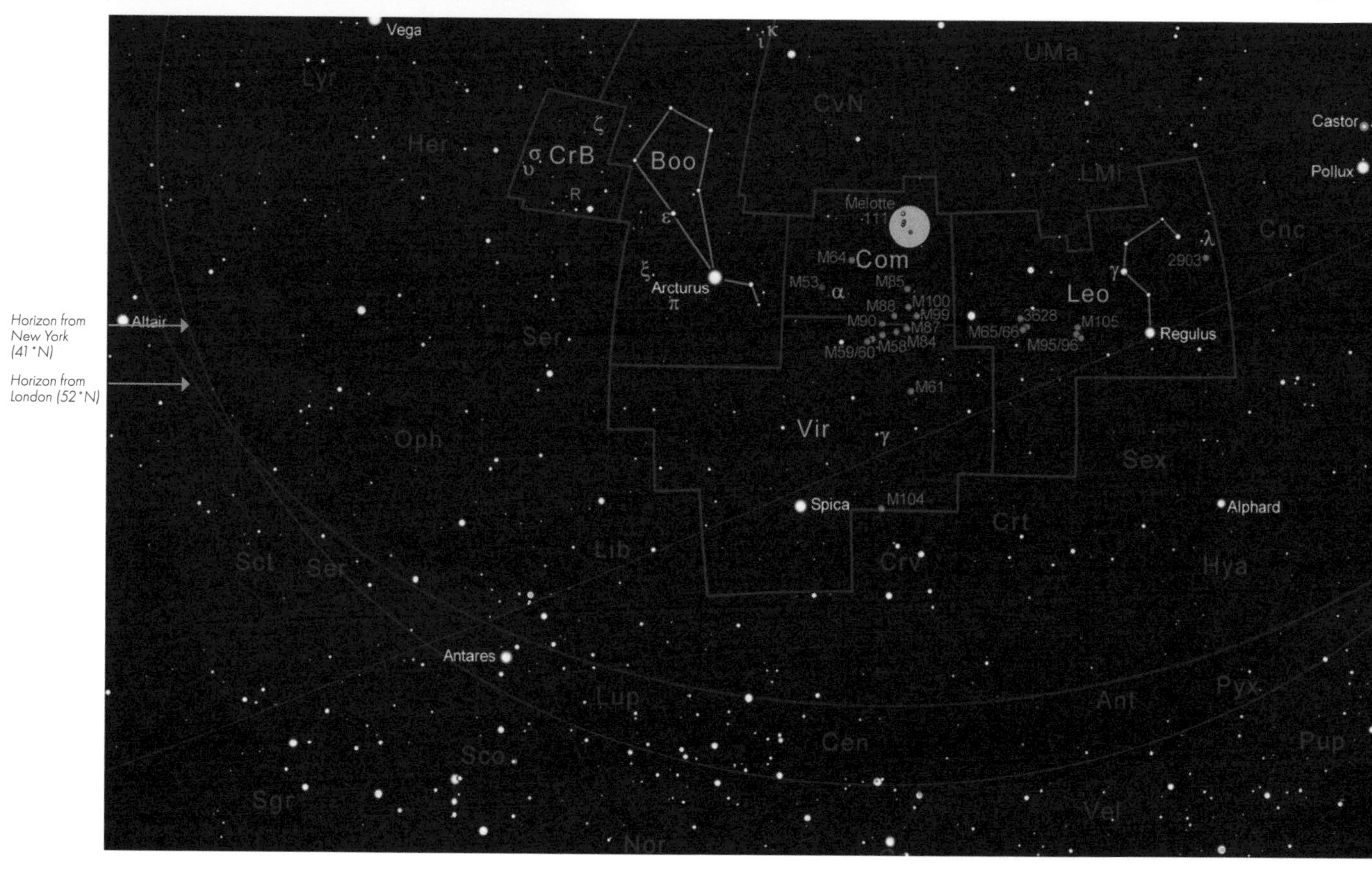
Horizon from New York (41°N)
Horizon from London (52°N)
Vega
Altair
Antares
Arcturus
Spica
Regulus
Alphard
Castor
Pollux
CrB
Boo
Com
Vir
Leo
Melotte 111
M64
M53
M85
M100
M99
M88
M90
M87
M84
M58
M59/60
M61
M104
M65/66
3628
M105
M95/96
2903

LEO

LEO / LEONIS

Highest at midnight: early March

Leo dominates its immediate neighborhood. It is easy to find, with the prominent Sickle asterism in the west and a bright triangle of stars forming its tail in the east.

Alpha Leo (Regulus) at the base of the Sickle is a bright double star whose components of magnitudes 1.4 and 7.7 can be easily separated through a small telescope. Being located less than half a degree from the ecliptic, Regulus is often occulted by the Moon. Gamma Leo is another charming double, made up of a pair of yellow stars of magnitudes 2.3 and 3.6.

A number of bright galaxies lie in the belly of Leo. M65 and M66, a pair of ninth-magnitude spiral galaxies, are less than half a degree apart; just to their north lies a fainter edge-on galaxy, NGC 3628. All three can be viewed in a low- power field through a 100mm telescope. Another notable triplet of galaxies lies a short distance to the west and comprises M95, M96 and M105. M95 is face-on with a bright nucleus, while M96 is a rather homogenous round smudge devoid of a nucleus. M105 is a cigar-shaped elliptical blur. The easily located spiral NGC 2903 lies one and a half degrees south of Lambda Leo, and shows some blotchiness through a small telescope.

Galaxies M65 and M66, observed by the author using a 300mm reflector (two telescopic fields of view combined); also visible are the fainter galaxies NGC 3628 (lower right) and NGC 3593 (far left).

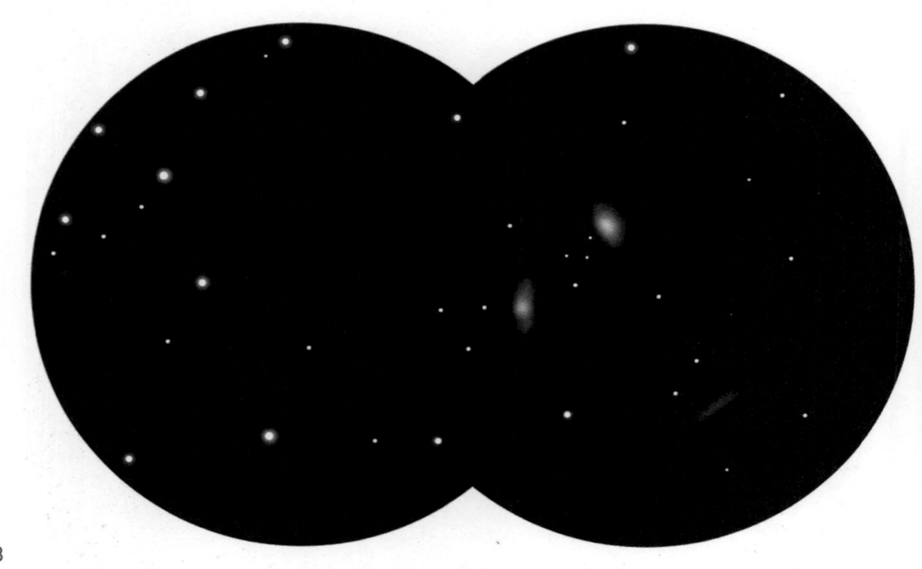

COMA BERENICES

COM / COMAE BERENICES

Highest at midnight: early April

Coma is an inconspicuous little constellation made up of around 30 faint stars on the border of naked-eye visibility.
The eye's attention is drawn to an extensive misty patch,the Coma Star Cluster (Melotte 111) containing around 40 stars (mostly below naked-eye brightness) spread over a five-degree wide area.

Hundreds of stars within M53, a bright globular cluster with a concentrated core, can be viewed through a 200mm telescope at high magnification. The cluster is easy to locate, just one degree northeast of Alpha Com.

All the brighter galaxies of Coma lie within the Coma-Virgo Cluster, a vast assembly of galaxies around 50 million light years away.

Probably the most spectacular is the Black Eye Galaxy (M64), a bright galaxy whose prominent silhouetted dark lane is easily visible through a 150mm telescope.
There are many more Coma galaxies easily visible through small telescopes, including M85, M88, M99 and M100.

The Black Eye Galaxy in Coma, imaged with a 127mm refractor and astronomical CCD camera.

VIRGO

VIR / VIRGINIS

Highest at midnight: mid-April

Stretching large and broad along the celestial equator, Virgo, the sky's second biggest constellation, occupies almost 1,300 square degrees of sky. Once the bright **Alpha Vir (Spica)** is located,
the other stars making up its pattern can be traced fairly easily.
To give the stargazer an idea of scale, two oustretched hand widths, with Spica beneath the intersection of the thumbs,
will just about cover the constellation's main body.

Gamma Vir (Porrima) is a famous double star of equal components, both white stars shining at magnitude 4.6. The pair orbit one another in a period of around 170 years,
and were closest in 2008 when they were only separable at high magnification using a 250mm telescope. In 2020 a good 60mm
telescope will be able to resolve them,
and they will be at their widest in 2080, when any small telescope will split them.

All of the bright galaxies in Virgo belong to the Coma-Virgo Cluster. Most of them
lie in the northwest of the constellation, and include M58, M59, M60, the Swelling
Spiral (M61), M84, M87 and M90. In the southwest of Virgo can be found the Sombrero (M104), a bright eighth-magnitude edge-on galaxy whose nuclear bulge rises smoothly on either side of its spindly spiral arms.
A dark dust lane running through the centre of M104 is visible through a 300mm telescope.

The Sombrero Galaxy in Virgo, observed by the author with a 300mm reflector.

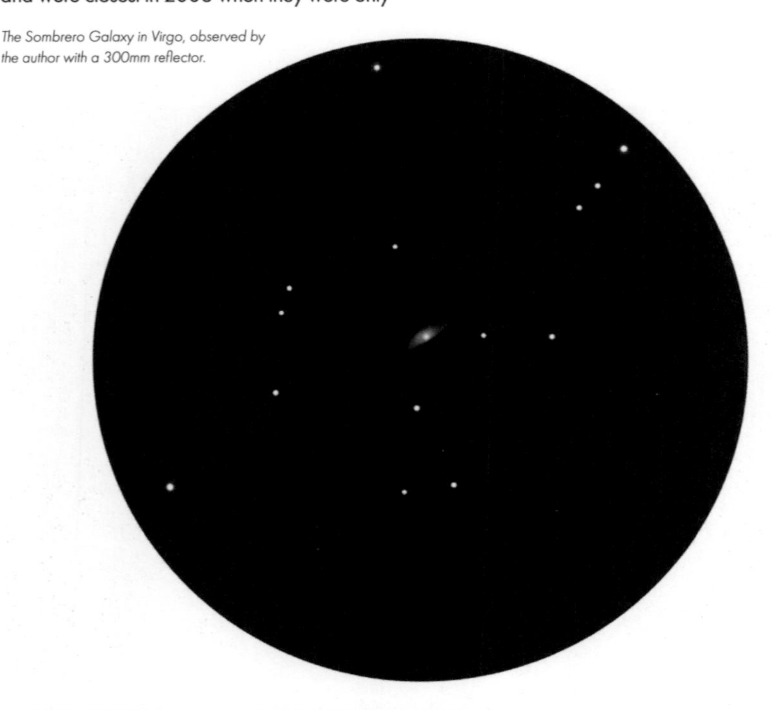

BOÖTES

BOO / BOÖTIS
Highest at midnight: early May

The brightest star in Boötes, the orange giant Alpha Boo (Arcturus) can be located by tracing a slightly curving line from the tail of Ursa Major. Many stargazers trace out the shape of Boötes by imagining a large kite, with Arcturus at its tail end and Beta Boo at its apex; the image is completed by a trail of stars west of Arcturus, representing the kite's trailing ribbon.

Although Boötes is a large constellation, it contains no bright deep-sky objects. It makes up for this with a number of fine double stars. Epsilon Boo is a lovely close double with an orange magnitude 2.5 primary and a blue magnitude 4.6 companion, and requires a 100mm telescope at x100 to resolve well.

Doubles that are easily visible through a 60mm telescope include Iota Boo, a wide double of magnitudes 4.8 and 8.3; Kappa Boo, an easy double of magnitudes 4.5 and 6.6; Pi Boo, magnitudes 4.5 and 5.8; and Xi Boo, a beautiful yellow - orange double of magnitudes 4.7 and 7.

Arcturus (Greek, meaning 'bear keeper'), the sky's fourth brightest star.

CORONA BOREALIS

CRB / CORONAE BOREALIS
Highest at midnight: late May

A small but delightful constellation, the main pattern of Corona Borealis is made up of a semicircle of seven bright stars.

Several doubles are easily separable through small telescopes, including Zeta CrB (magnitudes 5 and 6), Nu CrB (a wide fifth-magnitude pair) and Sigma CrB (magnitudes 5.6 and 6.6). R CrB is a dwarf nova cataclysmic variable star whose minima occur randomly. Mostly it shines around magnitude 6, but it can suddenly dim by up to eight magnitudes and can remain faint for many weeks.

Boötes and Corona Borealis

NORTHERN SUMMER STARS
(midnight, 1 July)

While gaining in one hand, the night sky observer loses in the other, as shirtsleeves observing conditions are accompanied by shorter, lighter nights. All the familiar winter constellations are below the horizon, although it is still possible to glimpse Capella in Auriga, a circumpolar star, low above the northern horizon.

Ursa Major has rotated around to the northwest,the Saucepan poised as if to pour its cosmic contents on faint Lynx below. Following the curve of the bear's tail takes us to Arcturus, which is slowly dropping in the western skies. The galactic feast begun in spring is coming to an end as Coma Berenices and Virgo sink into the west, only to be replaced with a sumptuous sampling of our own home Galaxy, the Milky Way.

Hercules, Corona Borealis and Ophiuchus – all lovely constellations with a wonderful selection of bright celestial gems – precede the famous Summer Triangle asterism made up of Vega in Lyra (now near the zenith), Deneb in Cygnus and Altair in Aquila. From the low southern constellations of Sagittarius and Scorpius, arching high through the Summer Triangle and into Cassiopeia and Perseus, can be seen the most spectacular section of the Milky Way visible from the northern hemisphere. When viewed from a dark-sky site, unhindered by light pollution, the unaided eye can discern remarkable detail along this section of the Milky Way. A prominent dark rift, caused by interstellar gas and dust seen in silhouette against distant stars in our Galaxy, runs through the Northern Cross asterism in Cygnus; further south, a seemingly detached section of the Galaxy known as the Scutum Star Cloud gives the northern hemisphere observer some idea of what the Magellanic Clouds look like to those living in the southern hemisphere.

Andromeda and the Square of Pegasus asterism, floating on an area of sky known to the ancients as the Water – containing the constellations of Pisces, Aquarius and Capricornus the Sea Goat – rise in the east. Cutting its lowest path across the skies for northern hemisphere observers, the ecliptic runs from faint Pisces in the southeast, through Sagittarius, low in the south, to Virgo in the west; at its highest the ecliptic is barely a hand's width above the southern horizon. Our Galaxy's central hub, located in Sagittarius, is masked by dark interstellar clouds viewed in silhouette, while a good sprinkling of beautiful bright nebulae can be seen in the vicinity.

Northern summer sky, looking due south (east at left, west at right) from the horizon to the zenith. The horizon lines for London (52°N) and New York (41°N) are marked, as well as the ecliptic. The chart is relevant for 1 May (4am), 1 June (2am), 1 July (midnight), 1 August (10pm) and 1 September (8pm).

Horizon from New York (41°N)

Horizon from London (52°N)

HERCULES

HER / HERCULIS

Highest at midnight: early June

A large constellation, identifiable by the **Keystone** asterism made up of four of Hercules' brighter stars – **Pi Her, Eta Her, Epsilon Her** and **Zeta Her**. These stars, combined with **Beta Her** and **Delta Her** further south, make up a familiar big butterfly-shaped asterism. The constellation stretches considerably further north, south and east of this, making Hercules the sky's fifth largest constellation.

Alpha Her is a splendid red giant varying between magnitudes 3 and 4; a small telescope will show that it has a close green-coloured companion of magnitude 5.4. Gamma Her and 95 Her are also nice telescopic doubles.

The Great Globular Cluster (M13) is the northern sky's brightest example of its type. Lying just two and a half degrees south of Eta Her, M13 is easy to locate – indeed, it is faintly visible with the unaided eye. Binoculars show it to be an extensive fuzzy patch measuring around half the apparent diameter of the full Moon. Viewed through a 150mm or larger telescope, the cluster is an amazing sight; the brightest of its 300,000 outlying stars can be resolved, and these appear to be arranged in several distinct radial lines. Hints of darker lanes can be discerned within the cluster's outer regions; photographs don't show these features well, but our perception through the eyepiece produces a different impression.

M92 is probably the northern sky's second most beautiful globular cluster. Located north of the Keystone, it receives less attention than its brighter sibling, but it is in many ways just as spectacular. Its outer stars can be resolved through a 150mm telescope, and the cluster is smaller and more compact and spherical than those of M13.

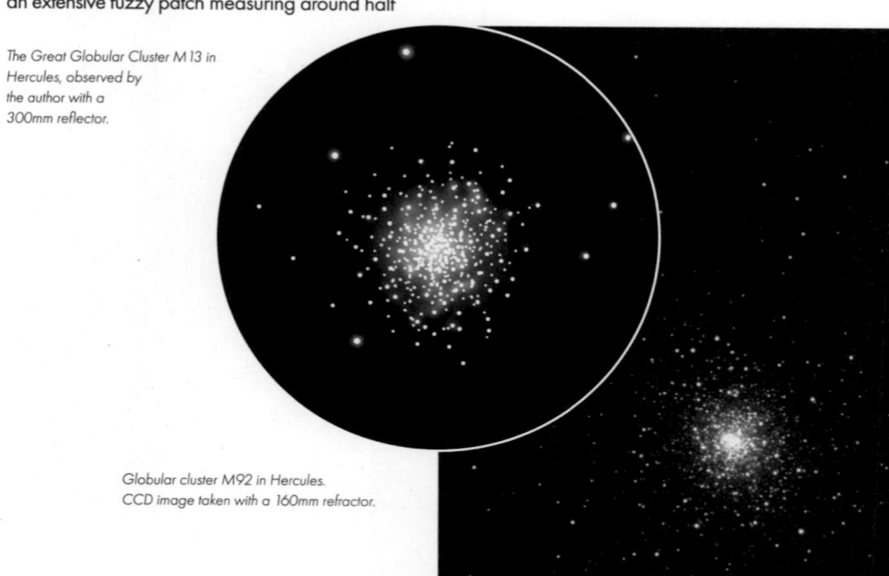

The Great Globular Cluster M13 in Hercules, observed by the author with a 300mm reflector.

Globular cluster M92 in Hercules. CCD image taken with a 160mm refractor.

OPHIUCHUS

OPH / OPHIUCHI

Highest at midnight: mid-June

A large constellation that extends well south of the celestial equator, Ophiuchus's main stars are of the second and third magnitude. Its full outline can be hard to trace from northern climes, since it extends to 30 degrees south.

The multiple star Rho Oph makes a great high-magnification sight. It consists of a magnitude 4.6 primary and close magnitude 5.7 partner, plus two more widely separated outlying stars of seventh magnitude. Another double worthy of note, 70 Oph, comprises components of magnitudes 4.2 and 6, which orbit each other in a period of 88 years; the pair will be at its widest in 2025.

North of Beta Oph lies IC 4665, a scattered open cluster visible with the naked eye as a misty patch, best viewed through binoculars.
Several degrees to the east lies Barnard's Star, a ninth-magnitude red dwarf, under six light years distant.

Ophiuchus is rich in globular clusters, seven of the brightest appearing in Messier's list (see Introduction): M9, M10, M12, M14, M19, M62 and M107. The brightest of these, M10 and M12, can be resolved through a 200mm telescope.

The Black Eye Galaxy in Coma, imaged with a 127mm refractor and astronomical CCD camera.

LYRA

LYR / LYRAE

Highest at midnight: early July

The compact constellation of Lyra is one of the best-known constellations of northern summer skies. Its western margin is clipped by a rich, wide section of the Milky Way. **Alpha Lyr (Vega)** is the fifth brightest star in the sky, and the second brightest in the northern celestial hemisphere. Vega is 25 light years away, measures more than three times the diameter of the Sun and is 61 times brighter. Vega was the first star ever photographed, back in July 1850. In 12,000 years' time, precession of the Earth's axis will have moved the north celestial pole near Vega.

Beta Lyr is a beautiful double star, the primary being a white variable star of magnitude 3.3–4.4; its companion a blue magnitude 7.2, Delta Lyr. The Double-Double Star (Epsilon Lyr) is a famous multiple. The main pair is wide enough to be separated with binoculars, while each of these is a close telescopic

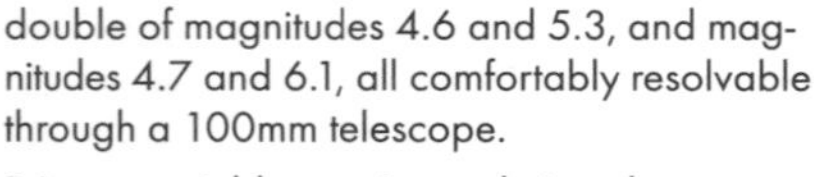

double of magnitudes 4.6 and 5.3, and magnitudes 4.7 and 6.1, all comfortably resolvable through a 100mm telescope.

R Lyr, a variable star, is a red giant that pulsates between magnitudes 3.9 and 5 every six to seven weeks. Some distance to its east lies RR Lyr, a special type of pulsating variable, which varies between magnitudes 7.1 to 8.1 in 13.6 hours.

The Ring Nebula (M57) is perhaps the best-known planetary nebula. Easily found midway between Beta Lyr and Gamma Lyr, M57 is rather small; binoculars show it as an almost star-like point, but through a telescope at a high magnification it resembles a sharply defined luminous ring. The brightest of Lyra's other deep-sky delights is the globular cluster M56. Many of its stars are resolvable through a 200mm telescope, and are set in a lovely rich galactic starfield.

The Ring Nebula in Lyra, imaged with a 200mm SCT and astronomical CCD camera.

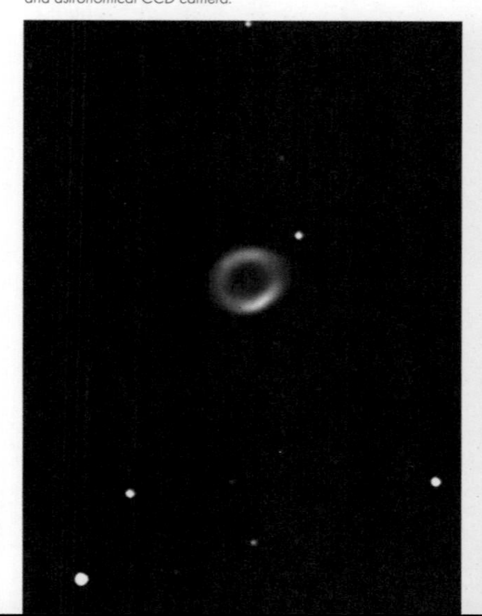

Globular cluster M56 in Lyra, imaged with a 160mm refractor and astronomical CCD camera.

AQUILA

AQL / AQUILAE

Highest at midnight: mid-July

Aquila, a medium-sized constellation, straddles the celestial equator.
Alpha Aql (Altair), the most southerly star of the prominent **Summer Triangle**, is one of our nearest stellar neighbours, lying just 17 light years away. Aquila contains two nicely coloured double stars, easily visible through small telescopes: **15 Aql**, a magnitude 5.4 orange star with a lilac magnitude 7 companion; and **57 Aql**, a sky-blue magnitude 5.7 primary with a magnitude 6.5 companion.

Aquila contains a fair scattering of faint planetary nebulae, but its deep- sky showpiece is NGC 6709, an open cluster made up of around 30 fairly bright stars, a number of which are arranged in loose chains.

Star cluster NGC 6709 in Aquila, observed by the author with a 100mm refractor.

CYGNUS

CYG / CYGNI
Highest at midnight: early August

Flying south along the Milky Way, the Swan is a wonderful constellation set against the bright, rich starry background of the Milky Way.
The stargazer can spend literally hours scanning this area through binoculars, sweeping along magnificent starfields and spending a while searching for some
of the more elusive deep-sky quarry on offer. As well as being one of the three bright stars that make up the Summer Triangle asterism, Alpha Cyg (Deneb) is the top star of the Northern Cross asterism.
At the foot of the cross, near the southwestern border of Cygnus, is Beta Cyg (Albireo), one of the most beautiful coloured double stars. A small telescope will easily resolve the companion to its golden magnitude 3.1 primary, a steely blue star of magnitude 5.1. Omicron Cyg is another wonderful coloured double, separable through binoculars, made up of an orange magnitude 3.8 primary and
a sea-green magnitude 4.8 companion. Closer scrutiny will reveal another companion to the primary, a blue magnitude 7 star.

The red giant Chi Cyg, a Mira-type variable star, can reach the third magnitude at its brightest, although it normally shines well below naked-eye visibility, dropping to as low as the
fourteenth magnitude. Its period is The red giant Chi Cyg, a Mira-type variable star, can reach the third magnitude at
its brightest, although it normally shines well below naked-eye visibility, dropping to as low as the fourteenth magnitude. Its period is made up of
around a dozen fairly bright stars and many fainter ones, and makes a lovely low-magnification telescopic sight.

The Blinking Planetary (NGC 6826)
lies east of Theta Cyg. It appears small and well defined, with a slightly blue tinge,
its central star visible through a 150mm telescope. Its outer shell appears to blink off only as the observer looks directly at the object, while its bright central star remains visible.

Double star Albireo, a beautiful contrasting pair of blue and gold set amid the Milky Way in Cygnus. Observation by the author with a 150mm refractor.

The North America Nebula in Cygnus, imaged with an astronomical CCD camera.

Spread across a portion of the Milky Way in southern Cygnus, the Veil Nebula is a supernova remnant, the brightest parts of which, NGC 6992, may be discerned through big binoculars from a dark site.

Big binoculars will also reveal the North America Nebula (NGC 7000), appearing as a wedge-shaped brightening of the Milky Way (considerably larger than the apparent diameter of the full Moon) to the east of Deneb. Having such a large area and a low surface brightness, it is elusive at higher magnifications through a telescope.

The Veil Nebula in Cygnus, imaged with an astronomical CCD camera.

NORTHERN AUTUMN STARS (midnight, 1 October)

After the short, light nights of summertime, autumn (fall) brings with it a dramatic change of celestial scenery. Ursa Major now paces the northern horizon, the Plough swinging low, while the W of Cassiopeia climbs high overhead, cut through by the great frieze of the Milky Way, which arches across the sky from east to west.

Hercules begins to step down towards the northwestern horizon after sunset, followed by Vega, which leads the Summer Triangle's descent. The Northern Cross is standing upright, or, if the traditional constellation image is pictured, the swan plummets head downwards towards the horizon.

While one hero departs in the west, another commences its ascent in the east after sunset. Orion has just about cleared the horizon, his brilliant and easily recognizable hunter's form pursuing red-eyed Taurus.
Walking hand-in-hand with each other, the twins Castor and Pollux in Gemini stroll into the autumn (fall) skies; while Auriga's chariot thunders above them,
its alpha star Capella shining as a prominent beacon high above the eastern horizon.

Looking south, a large expanse of sky is filled with relatively faint stars and ill-defined star patterns. Although these autumn (fall) constellations may lack the immediate visual impact of many other parts of the sky,
they are, nonetheless, packed full of interest. Our view into Sculptor, near the southern horizon, is at right angles to the plane of the Milky Way, and we're peering into intergalactic depths unobstructed by most of the dust and gas of our home Galaxy.

Given a clear horizon, the bright Fomalhaut in Piscis Austrinus flashes its presence in the celestial Water that runs across the southern sky from Eridanus in the southeast, through Cetus,
Pisces and Aquarius, into Capricornus in the southwest. Climbing in the east, orange Aldebaran is preceded by the Pleiades in Taurus; to its west is the familiar little pattern of Aries, and next
to it the ill-defined pattern of stars making up Pisces, traceable around the bottom left corner of the Square of Pegasus. Andromeda and Triangulum soar high, their boundaries containing the two most distant objects visible with the unaided eye – the Great Spiral in Andromeda and the Pinwheel Galaxy.

Northern autumn (fall) sky, looking due south (east at left, west at right) from the horizon to the zenith. The horizon lines for London (52°N) and New York (41°N) are marked, as well as the ecliptic. The chart is relevant for 1 August (4am), 1 September (2am), 1 October (midnight), 1 November (10pm) and 1 December (8pm).

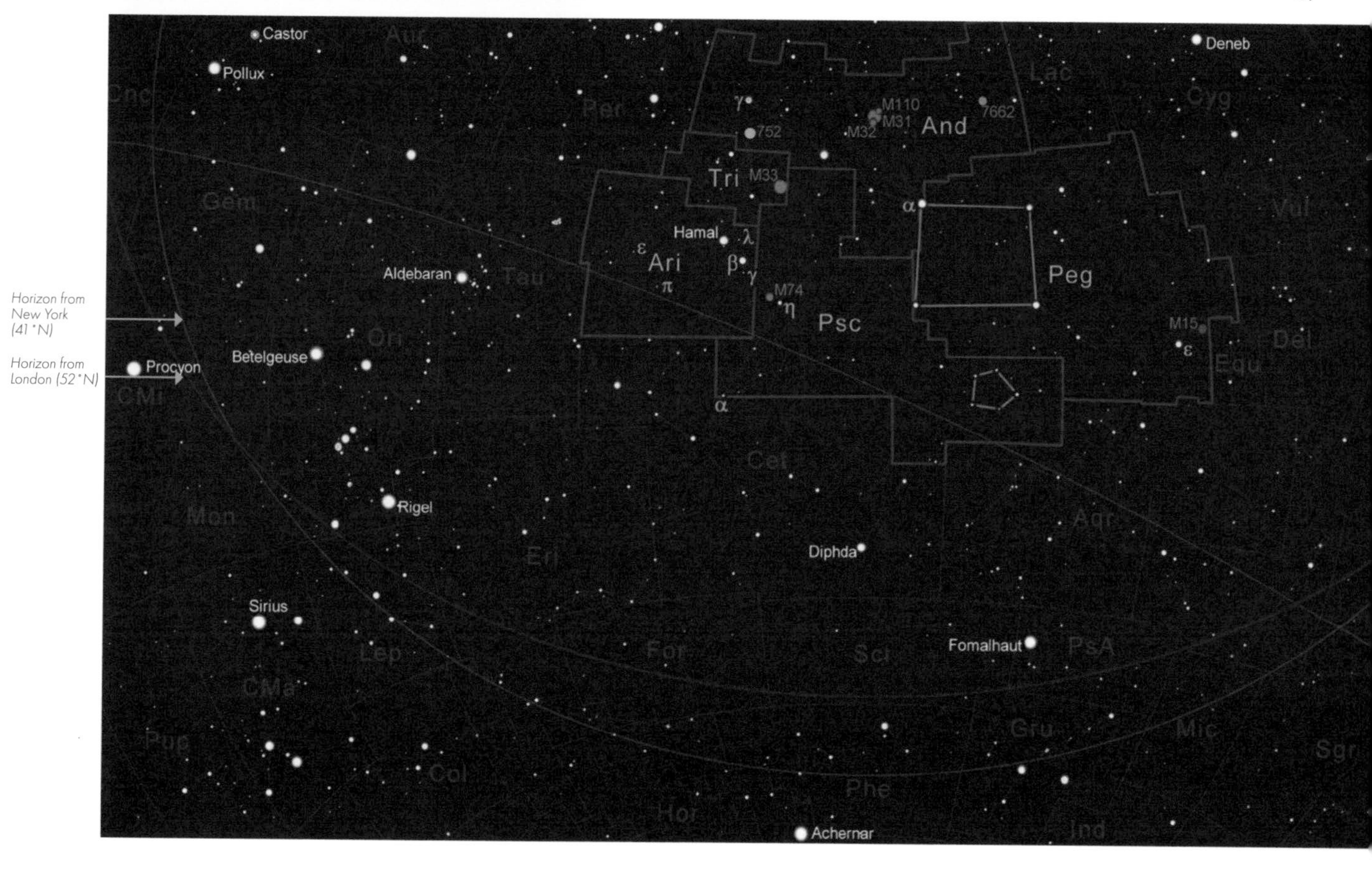
Horizon from
New York
(41°N)
Horizon from
London (52°N)
Castor
Pollux
Procyon
Betelgeuse
Aldebaran
Rigel
Sirius
Hamal
Ari
Tri
M33
752
M110
M31
M32
And
7662
Deneb
Peg
Psc
M74
M15
Diphda
Fomalhaut
Achernar
Gem
Tau
Ori
Per
Aur
Mon
Eri
Cet
Lep
CMa
CMi
For
Scl
PsA
Aqr
Gru
Phe
Hor
Col
Pup
Ind
Mic
Sgr
Equ
Del
Vul
Cyg
Lac

PEGASUS

PEG / PEGASI

Highest at midnight: early September

Pegasus is a wide constellation, easy to locate because the prominent Square of Pegasus asterism takes up much of its eastern parts – but note that the top left star actually belongs to Andromeda. Seeing how many stars are visible within the square is a good test of the darkness of your observing site; if you can spot six stars, then you have a nice dark site.

Epsilon Peg is a lovely wide double star comprising an orange magnitude 2.4 primary and a blue magnitude 8.4 companion. Around four degrees to its northwest lies M15, a nice globular cluster bright enough to be seen through binoculars; its outer stars are resolvable through a 150mm telescope.

The North America Nebula in Cygnus, imaged with an astronomical CCD camera.

ANDROMEDA

AND / ANDROMEDAE

Highest at midnight: late September

Andromeda covers a width of sky about equal to two outstretched hands placed thumb to thumb.
Although it's not a particularly brilliant constellation, Andromeda is easily found, since its main star, Alpha And (Alpheratz) is the top left-hand star of the Square of Pegasus asterism. Eastward from Alpheratz, Andromeda's brighter stars trace out a widening cutlass-blade shape.

Gamma And (Almach) is a lovely double star with gold and blue components of magnitudes 2.3 and 4.8, easily separable through a small telescope.
The fainter component also has a close blue magnitude 6.6 companion which will become resolvable through a 200mm telescope around 2020.

Andromeda is best known for being home to the Great Andromeda Galaxy (M31), the largest member of the Local Group of galaxies, more than 2.5 million light years away. It is easy to locate and can be seen without difficulty with the unaided eye from dark suburban sites. We view M31 from an angle of around 30 degrees above its plane, so that it is somewhat foreshortened.
Through binoculars it appears as a bright misty oval around half a degree wide, and from dark sites it stretches yet further across the field of view. A 200mm telescope will reveal hints of structure within the galaxy, including a prominent dark lane and a suggestion of knottiness (large nebulae) in its spiral arms.

Nearby are its small satellite galaxies, M32 and M110, both visible as tiny condensed blobs through 80mm binoculars. M32, the brighter of the pair, lies around half a degree south of M31's centre, while M110 lies around a degree to
the northwest.

NGC 752 is a large open cluster made up of more than 60 faint stars spread pretty evenly over an area larger than the full Moon. The cluster is visible as a
misty patch through binoculars, and its individual stars are resolvable through a 100mm telescope. The Blue Snowball (NGC 7662) is a bright, ninth-magnitude planetary nebula. It is visible through small telescopes as a fuzzy blue spot, and delightful to view through larger instruments, though its colour may not be so apparent at high magnifications.

Comparison between an astronomical CCD image and a visual observation of the Great Spiral in Andromeda (plus its companion galaxies M32 and M110). Field of view is two degrees (four times the apparent width of the Moon).

PISCES

PSC / PISCIUM

Highest at midnight: early October

Lying to the immediate south and east of the Square of Pegasus, Pisces is one of the largest of the 12 Zodiacal constellations.

Its traditional outline, comprising a series of faint stars, is only traceable from dark sites. Pisces' brightest star, Alpha Psc, lies in the far southeastern corner of the constellation. Through a 100mm telescope it can be resolved as a close double of magnitudes 4.2 and 5.2. In the western corner of Pisces, the well-known asterism of the Circlet is made up of seven stars – a challenge to spot with the unaided eye from an urban site.

The Phantom (M74), a face-on spiral galaxy, is Pisces' brightest deep-sky object. It can be found a little more than one degree east of Eta Psc, and appears as a sizeable round smudge with a bright, well-defined nucleus through a small telescope.

The Phantom (so called because of its low surface brightness), a face-on galaxy M74 in Pisces, imaged using a 127mm refractor and astronomical CCD camera.

TRIANGULUM

TRI / TRIANGULI
Highest at midnight: late October

Wedged between Aries and Andromeda, Triangulum is one of the sky's smallest and least prominent constellations. It is made up of three faint stars that form an elongated triangle. Insignificant though Triangulum appears, it hosts one of the nearest galaxies – the **Pinwheel Galaxy (M33)** – a face-on spiral some 2.7 million light years distant, just visible with the naked eye from a dark site. This galaxy has a low surface brightness, so although it may be seen through binoculars, it may be missed using a telescope with a higher magnification.

The Pinwheel Galaxy in Triangulum, a face-on, low surface object. Imaged with a 127mm refractor and astronomical CCD camera.

ARIES

ARI / ARIETIS
Highest at midnight: early November

Aries, the smallest constellation of the Zodiac, can be identified by the small pattern of its brighter stars – **Alpha (Hamal) Ari, Beta Ari** and **Gamma Ari** – which lie some distance west of the Pleiades in neighbouring Taurus. The Sun, Moon and planets are frequent visitors to the southern part of Aries, since this is where a short section of the ecliptic lies.

Gamma Ari is one of the best identical stellar duos in the sky; a double of white magnitude 4.6 stars, easily visible through small telescopes and looking like a pair of glowing eyes. Lambda Ari is another wide double, with a white magnitude 4.8 primary and a yellow magnitude 7.3 companion. More challenging doubles in Aries include Epsilon Ari, a close pair of white stars of magnitudes 4.6 and 5.5, just separable with a 100mm telescope; and Pi Ari, a blue magnitude 5.2 star with a close yellow magnitude 8.5 companion, separable with a 60mm telescope. The constellation harbours no conspicuous deep-sky objects, but ardent deep-sky hunters may attempt observing a smattering of twelfth-magnitude galaxies that lie within Aries' borders.

Gamma Ari, a twin double star, observed through a 100mm refractor.

THE SOUTHERN STARS

In this small section, we are showing the remaining zodiac constellations that can be seen in the Southern Hemisphere.

LIBRA ♎

LIB / LIBRAE
Highest at midnight: mid-May

One of the smallest and least conspicuous of the Zodiacal constellations, Libra's main stars form a quadrilateral that straddles the ecliptic. It is located northwest of Scorpius, but it is difficult to see with the unaided eye from a city.

Alpha Lib (Zubenelgenubi), a wide double separable in binoculars, consists of a skyblue magnitude 2.7 star with a white magnitude 5.2 companion.
Beta Lib is the brightest of Libra's stars and displays an uncommon green hue, the colour being particularly notable through binoculars.

The False Comet Cluster in Scorpius, imaged with a 250mm reflector and astronomical CCD camera.

The Trifid Nebula in Sagittarius, imaged with a 250mm reflector and astronomical CCD camera.

SCORPIUS

SCO / SCORPII

Highest at midnight: early June

Almost overhead in southern winter skies, Scorpius is a spectacular constellation highlighted by the distinctly orange Alpha Sco (Antares). Although low to the horizon when seen from northern temperate locations, the collection of bright stars in the tail of Scorpius near Antares are a familiar sight to northern temperate stargazers (though somewhat dimmed by atmospheric murk). Through a telescope, the fainter blue companion to Antares (an orange supergiant) may be glimpsed through quite a small instrument, although the brilliance of Antares hinders its visibility.

West of Antares, the misty patch of globular cluster M4 can be seen with the unaided eye. A 100mm telescope will resolve it.

SAGITTARIUS

SAG / SAGITTARII

Highest at midnight: early July

Sagittarius is justifiably considered by many stargazers – especially those in the southern hemisphere, who see it climb high overhead in winter – to be the grandest constellation of them all. It is crossed by a broad swathe of the Milky Way, and beyond it is the centre of our Galaxy, positioned near the constellation's far western border. Sagittarius is full of deep-sky splendours.

No fewer than 15 Messier objects (see Introduction) lie within Sagittarius, most of which appear against the backdrop of the Milky Way in the western half of the constellation, and here can be found the loveliest collection of these.

The Milky Way in Scorpius and Sagittarius, imaged from La Palma with a driven digital SLR.

CAPRICORNUS

CAP / CAPRICORNI

Highest at midnight: early August

A fairly small Zodiacal constellation, and quite an obscure one too, Capricornus comprises a collection of third- and fourth-magnitude stars arranged in a broad south-pointing arrowhead, 'roof' or 'tent' pattern when viewed from the southern hemisphere.

Alpha Cap is a close naked-eye double comprising Alpha 2 Cap, a yellow giant of magnitude 3.7 and Alpha 1 Cap, an orange supergiant of magnitude 4.3. The pair is an easily separated line-of-sight double, the components lying at 109 and 887 light years away respectively. Each of these stars is itself a wide double with faint companions, the dimmest visible through a 200mm telescope. Beta Cap is a nice coloured double with a golden magnitude 3 primary and a sky-blue magnitude 6.1 partner, wide enough to be seen through binoculars.

Capricornus's brightest deep-sky delight is M30, a middling sized, fairly bright globular cluster whose outer regions can be resolved through a 200mm telescope.
Residing in the constellation's southwestern corner is its brightest galaxy, NGC 6907, a beautiful twelfth-magnitude face-on barred spiral, best viewed through telescopes larger than 250mm.

Double star Beta Capricorni, observed by the author using a 200mm SCT.

AQUARIUS

AQR / AQUARII

Highest at midnight: late August

Aquarius is a large constellation whose main pattern comprises a widely spaced collection of second- and third-magnitude stars lying south of the celestial equator. Immediately east of Alpha Aqr, a small but striking asterism often called the Propeller or the Water Jar comprises a pattern of four stars, whose central star, Zeta Aqr, is a binary with twin white components of magnitudes 4.3 and 4.4.

The system is slowly opening up to us; it can be resolved through a good 80mm telescope, and by the end of the 21st century it will be an easy double through any small telescope.

Three different types of deep-sky object – a planetary nebula, an open cluster and a globular cluster, all visible in the same field of view at low magnifications – can be found in the western reaches of Aquarius. Less than one and a half degrees west of Nu Aqr lies the Saturn Nebula (NGC 7009), a superb planetary nebula, visible through a 80mm telescope as an elliptical disk of a similar size to Saturn. Larger instruments will reveal the blue colour and some structure, including a narrow inner ellipse and two small lobes protruding from the nebula, giving it the appearance of Saturn with its rings presented edge-on to us. To its southwest is M73, a small open cluster shaped rather like the Propeller asterism. Nearby M72, a small ninth-magnitude globular cluster with a bright core, can be difficult to resolve using anything smaller than a 250mm telescope. M2, the brightest globular cluster in the region, is located less than five degrees north of Beta Aqr.

It can be seen through binoculars as a fuzzy patch, and a large number of its stars can be seen through a 150mm telescope.

In the far south of Aquarius, the Helix Nebula (NGC 7293) can be seen fairly easily through binoculars as a circular smudge. Although it has the largest apparent diameter of all planetary nebulae, it has a low surface brightness, so the best views are at low magnifications. Hints of its ring structure and some mottling may be glimpsed through a 250mm telescope.

The Saturn Nebula in Aquarius, observed by the author using a 300mm reflector.

PICTURE CREDITS

Anthony Ayiomamitis

5, 21(r), 23(r), 27, 33, 34, 41(l), 44(l), 46(r)

Peter Grego

2, 3, 4, 7, 9, 12, 13, 14, 15, 17, 19, 21(l), 23(l), 29, 31(t), 38, 40, 41(r), 43, 44(r), 46 (l), 47, 48(l), 51, 52, 75(r), 55(r), 58, 59

Nick Howes

22, 31(b), 49, 53(l)

NASA

11

NASA / STScl / Steinberg / Adam Block / NOAO / AURA / NSF

10

Public Domain

8

Peter Vasey

3, 4, 24, 25, 26, 30, 32, 35, 37, 39, 45, 48(r), 54, 55(l), 56, 57

Star symbols used throughout:
Denis M Moskowitz and Alec Finlay
http://www.suberic.net/~dmm/astro/constellations.html

This short book is taken from
The Star Book: How to Understand Astronomy by Peter Grego.